ATOMIC THUNDER

THE MARALINGA STORY

ELIZABETH TYNAN

NEWSOUTH

A NewSouth book

Published by
NewSouth Publishing
University of New South Wales Press Ltd
University of New South Wales
Sydney NSW 2052
AUSTRALIA
newsouthpublishing.com

National Library of Australia
Cataloguing-in-Publication entry

    Creator: Tynan, Elizabeth, author.
    Title: Atomic Thunder: The Maralinga story / Elizabeth Tynan.
    ISBN: 9781742234281 (paperback)
           9781742242446 (ebook)
           9781742247830 (ePDF)
    Notes: Includes index.
    Subjects: Nuclear weapons – Great Britain – Testing.
      Aboriginal Australians, Treatment of – South Australia – Maralinga.
      Radioactive pollution – South Australia – Maralinga.
      Cold War.
      Australia – Politics and government – 1945–1965.
      Great Britain – Politics and government – 1945–1964.
      Maralinga (SA).
Dewey Number: 994.05

Design Josephine Pajor-Markus
Cover design Blue Cork
Cover images Landscape at Maralinga, 2007: Wayne England/Wikimedia
Commons. John L Stanier at Maralinga in protective clothing, with a camera
also protected in a plastic cover [detail], c. 1950s: National Archives of
Australia A6457, P214.
Printer Griffin Press.

Extract from 'The Boy in the Bubble', p. 11: Words by Paul Simon. Music by Paul
Simon & Forere Motloheloa. © Copyright 1986 Paul Simon (BMI). All rights
reserved. International copyright secured. Used by permission of Music Sales
Limited.

ATOMIC THUNDER

ELIZABETH TYNAN is an academic at the James Cook University Graduate Research School in Queensland. A former journalist, she has a background in print and electronic media; previously she was a reporter and subeditor at the ABC and a correspondent for *New Scientist* magazine. She has also worked as a writer and editor at CSIRO, the Australian National University and the Australian Institute of Marine Science. Born in South Australia, she has long been fascinated by Maralinga and in 2011 completed a PhD on aspects of British nuclear testing in Australia. She is also a freelance science writer and editor, and co-author of the textbook *Media and Journalism: New Approaches to Theory and Practice*, now in its third edition, and of *Communication for Business*.

To the future.
May it learn something from the past.

Contents

Acknowledgments

As a bookish introvert, I enjoy nothing more than the challenge of living inside a big writing project. I have lived inside this one for quite a while. I haven't been entirely on my own, though. In fact, without the magnificent contribution of a number of people, there would be no book. I hasten to add that while I have been greatly assisted by some excellent individuals, if there are any errors in this book they are mine alone.

My incomparable mother, Rosemary Jennings, has been central to the creation of this book. Her experience as a history researcher, including work for the *Australian Dictionary of Biography*, honed her acute historical brain and great love of history. She is also naturally pedantic and has directed this superpower onto my work. She has picked up errors that I have not been able to see, and she has been a sounding board for my ideas. She is one of the few people who will allow me to talk at length about Maralinga without suddenly remembering that she has to be somewhere else. I am grateful to have been able to do that, because sometimes the ideas just want to come out, and having someone willing to receive them has been inexpressibly important to me.

I would also like to thank radiation scientists and Maralinga experts Dr Geoff Williams and Mr Peter Burns. My visit to Geoff and Peter at the Australian Radiation Protection and Nuclear Safety Authority in Melbourne in 2004 planted the seed of an idea that later became my PhD thesis and still later became this book. I never knew that the Maralinga story was so rich and fascinating and terrible until I spent those crucial, life-changing hours in the presence of such knowledgeable scientists. Both Geoff and Peter have been kind enough to read parts of the manuscript to check for factual accuracy.

Geoff also introduced me to Graeme Newgreen, who worked at Maralinga during the tests, and Graeme kindly read part of the text too.

Paul Malone was one of the intrepid investigative journalists who took on th Maralinga story when it became an important media event in the 1980s. His work with Howard Conkey at the *Canberra Times* revealed a complex story that he worked meticulously to uncover. He has generously given me access to his extensive archive of original documents relating to the nuclear tests in Australia, and I have drawn upon them gratefully and at length.

One of Australia's best journalists, Brian Toohey, broke the story about plutonium contamination in a series of stories in the *Financial Review* in 1978. He kindly answered my questions about his Maralinga reporting when I put them to him while researching my PhD. I have quoted those answers in this book as well, to help provide some insights into the era of uncovering that he did so much to initiate. I acknowledge the considerable contribution of the late Ian Anderson, a science journalist of great talent and influence who was taken too soon. His work in the early 1990s in uncovering the true extent of plutonium contamination caused by Vixen B at Maralinga was an object lesson in why investigative journalism is essential in a democracy. I was privileged to work with him briefly at *New Scientist*; he taught me so much. His widow, Dr Robin Anderson, generously gave me access to parts of Ian's personal archive in the early stages of my PhD research.

I gratefully acknowledge my employer, James Cook University. Part of the work involved in this book was carried out during a period of study leave in 2014. I particularly acknowledge the dean of graduate studies, Distinguished Professor Helene Marsh, who has always shown heartening and much-appreciated confidence in my abilities. Thank you also to linguists extraordinaire Professor Alexandra Aikhenvald and Professor Robert Dixon at James Cook University, who did the detective work that tracked down the origin of the word Maralinga.

Acknowledgments

This book had its genesis in the work I did for my doctorate. I would like to thank and acknowledge my supervisor at the Australian National University (ANU), Professor Sue Stocklmayer, and also Dr Will J Grant, both at the National Centre for the Public Awareness of Science. My PhD was a life-changing experience, and I thank them for their role in it.

My dear friend Susan Davies, who lives in New York and has been away for so long, will always be close to my heart. There truly is no friend like one's oldest friend. I am fortunate all-round in the quality of my friends. Special mentions to Melissa Lyne, Nicola Goc, George Roberts, Nadine Marshall, Marilyn Chalkley and Annie Warburton, who have all in a multitude of ways enriched my life.

My family is kind, loving and supportive, and I care for them deeply. With gratitude and love, I thank Dad (Frank), Inta, Meredith, Andrew, Narelle, Sophie and Alexander. As a long-time ANU employee, Dad knew some of the key players in this story and was able to share some tales. Also, my grandfather, Dad's dad, worked for a while with Len Beadell, and Dad has helped me source material about Beadell's exploits. I also thank my wider family – my delightful aunts, uncles and cousins – and mention in particular my dear Uncle Glen, who has always taken a keen interest in my Maralinga research. Thanks also to Brett.

An important marker of a robust democracy is ready access to a nation's documents. I have made extensive use of the National Archives of Australia and the National Archives of the United Kingdom and have always found the experience rewarding and, indeed, rather exciting. There's nothing like a set of old documents to get the blood racing. Sincere thanks to the staff at both archives for assisting me so ably.

I am indebted to Phillipa McGuinness at NewSouth Publishing, who saw promise in the Maralinga story and decided to take a chance on me. Thanks also to the always friendly and efficient Emma Driver, who has helped guide me through the process of becoming a NewSouth author. The term eagle-eyed barely begins to cover the talents of the editor Victoria Chance, who has been dogged and meticulous in editing this manuscript. Her highly professional work has made a huge difference to the quality of the final product. Thanks also to the proofreader Penny Mansley and the indexer Trevor Matthews, who have carried out their detailed work with admirable diligence.

I can't imagine life without the various animals who have filled my heart. I mention in particular Higgy, Minnie, Palmee, Samira, Ramona, Adelaide, Elvis, Lukey, Fred and Agnes. My heart still aches for those no longer with me, in particular Monty, Ava, Miranda, Bobby, Lily, Rosie, Wilfred and Wilma.

The Maralinga story is a vast, sprawling saga. This book is an attempt to provide a concise overview that will be of interest to the general reader, as well as offering a fresh perspective based upon years of analysis of the many diverse forms of evidence available. Many people have a profound stake in the events at all three test sites, most especially Indigenous people and service personnel (the 'nuclear veterans'). My book does not seek to delve into the fine detail of the grievances of either of these groups, not because their grievances are irrelevant or uninteresting, but simply because to do so would make this a different book altogether. I have instead sought to broaden the view to show Maralinga in its historical and scientific context. What an honour it is to write such a story.

Abbreviations

ABC	Australian Broadcasting Commission (from 1983, Corporation)
AERE	Atomic Energy Research Establishment (UK)
AIRAC	Australian Ionising Radiation Advisory Council
ALP	Australian Labor Party
ANU	Australian National University
ARL	Australian Radiation Laboratory
ARPANSA	Australian Radiation Protection and Nuclear Safety Agency
ASIO	Australian Security Intelligence Organisation
AWRE	Atomic Weapons Research Establishment (UK)
AWTSC	Atomic Weapons Tests Safety Committee
CRO	Commonwealth Relations Office
CSIR	Council for Scientific and Industrial Research
CSIRO	Commonwealth Scientific and Industrial Research Organisation
HER	High Explosive Research (UK)
IAEA	International Atomic Energy Agency
LRWE	Long Range Weapons Establishment
MARTAC	Maralinga Rehabilitation Technical Advisory Committee
MAUD	Military Application of Uranium Detonation (UK)
MEP	Maralinga Experimental Programme (UK)
RAAF	Royal Australian Air Force
RADSUR	Radiation Survey (UK)
RAF	Royal Air Force (UK)
TAG	Technical Assessment Group
TNT	trinitrotoluene

UK	United Kingdom
US	United States
USSR	Union of Soviet Socialist Republics

Measurements

During the period of the British nuclear tests, Australia used imperial measurements, and many of the quotes in the book reflect this.

    1 inch = 2.5 centimetres
    1 mile = 1.6 kilometres
    1 pound = 0.45 kilograms
    1 ton = 0.907 tonnes

Also, until February 1966, Australian currency was pounds, shillings and pence. At the time of the changeover, one Australian pound equalled two Australian dollars.

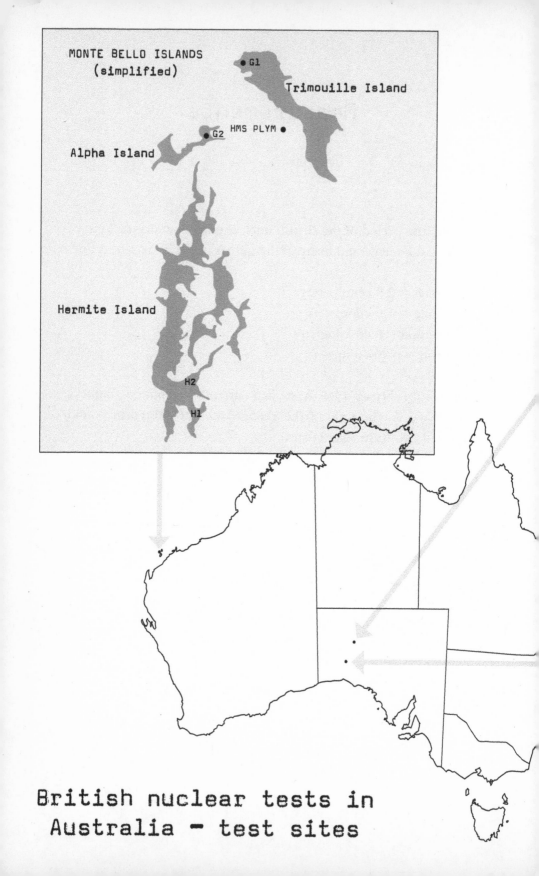

MONTE BELLO ISLANDS
(simplified)

● G1

Trimouille Island

● G2 HMS PLYM ●

Alpha Island

Hermite Island

H2

H1

British nuclear tests in Australia - test sites

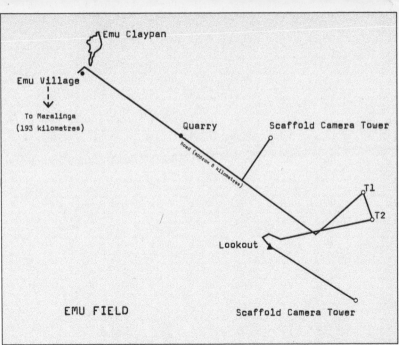

EMU FIELD

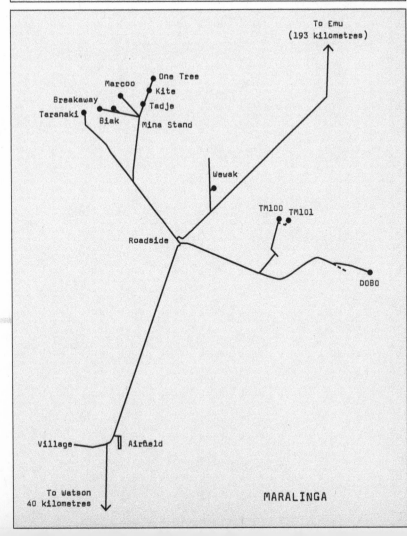

MARALINGA

Prologue

Maralinga. The name rolls easily off the tongue. It is a rather beautiful name, an Aboriginal word, but fittingly, given the colonialism at the heart of the Maralinga story, one not anchored in the place itself. The Indigenous people who lived in this part of South Australia for tens of thousands of years never spoke this word until it was transplanted there by white men. The name, from an extinct Aboriginal language called Garik, was officially adopted at a meeting of six Australian public servants and senior military personnel, the Research and Development Branch of the Commonwealth Department of Supply. At 10 am sharp on Wednesday 25 November 1953, long-time chief scientist for the department, the New Zealand–born Alan Butement, tabled it as the first order of business. He almost certainly got the name from anthropologists working in the Northern Territory, although the meeting minutes do not record that detail.

The new name met with the approval of the British 'nuclear elite', the top nuclear scientists from the Atomic Weapons Research Establishment (AWRE) at Aldermaston in southeast England. Charged with finding the right place to test British nuclear weapons, these men appropriated thousands of square kilometres of South Australian desert known to surveyors simply as X300. They turned a pristine Australian wilderness into one of the most contaminated places on earth in the pursuit of technological and geopolitical might for the United Kingdom (UK).

The nuclear tests started in October 1952 at Monte Bello Islands off the coast of Western Australia and moved briefly to a remote South Australian site called Emu Field in 1953. Even before they went to Emu, though, the scientists knew that it was not suitable for the expansive permanent location they wanted. Instead, Maralinga, not far to the south of Emu, was destined to be the final choice. A formal agreement to carry out atomic tests at Maralinga was signed by the British and Australian governments on 7 March 1956. The first major bomb tests got underway there six months later.

The word Maralinga means 'thunder' in Garik, a language once spoken by the people who lived around Port Essington. This short-lived British settlement, established in the early nineteenth century on the Cobourg Peninsula across from Darwin, today lies in ruins. Maralinga was one of a handful of Garik words recorded by anthropologists working in the territory; there are no known speakers today. Those who bound the word forever to the wildly beautiful red dust land in South Australia knew that it was exactly the right name. The thunder that rolled across the plains was an ominous sound that heralded a new leading player in a nuclear-armed and infinitely more dangerous world.

The British nuclear tests in Australia had their direct beginnings in the Manhattan Project. This secret wartime project created the atomic bombs dropped on the Japanese cities of Hiroshima and Nagasaki in August 1945, effectively ending the war in the Pacific. The project harboured atomic physicist spies, and their uncovering cleaved the alliance between Britain and the United States (US) that had produced the bombs. The British then turned their eyes towards the vast open spaces of Australia.

Indirectly, historical forces had long been conspiring to lead British scientists to the Australian outback. The British colonisation of Australia in the eighteenth century may well be the true starting point for this saga. The English explorer James Cook first planted the Union Jack on Australian soil in April 1770, during his epic scientific expedition. Soon after, the entire continent was absorbed into the British Empire, where it remained until 1901. This created a power differential in the relationship between the two lands. Even after Australia became a sovereign nation, strong echoes of its colonial past rang down through the generations, including the years when the British conducted nuclear tests on Australian territory between 1952 and 1963.

A subspecies of the colonialism that first claimed this island continent pervades this story. After World War II, as Britain's remaining colonies achieved independence one by one, its days as the world's biggest imperial power petered out. Colonialism as a broader force receded, but a new form emerged: nuclear colonialism. The term was coined recently – in 1992 – by the US anti–nuclear weapons testing activist Jennifer Viereck, who described it as 'the taking (or destruction) of other peoples' natural resources, lands, and wellbeing for one's own, in the furtherance of nuclear development'. The term – with its connotations of dominance and imperial superiority – fits the experience in Australia. When the call came from 'home', Robert Menzies, prime minister at the time, did not hesitate: Australian territory was immediately put at the disposal of the British, initially without any democratic niceties. In effect, the democratically elected prime minister of Australia decided to 'lend Australia to the United Kingdom' without the consent of its people. This, pointedly, was the first of the 201 conclusions of the Royal Commission into British Nuclear Tests in Australia, chaired by James McClelland, in the mid-1980s.

A phone call was all it took. The UK prime minister Clement Attlee rang Menzies in September 1950 after the British high commissioner in Canberra had passed on a top-secret message on 16 September. The message, from Attlee to Menzies, said in part, 'I am telegraphing to you now to ask first whether the Australian Government would be prepared in principle to agree that the first United Kingdom atomic weapon should be tested in Australian territory and secondly, if so, whether they would agree to our experts making a detailed reconnaissance of the Monte Bello Islands so that a firm decision can be taken on their suitability'. Menzies agreed without hesitation. The matter was not presented to Cabinet. The test date was to be sometime in 1952, as British scientists were scrambling to finalise construction of a workable nuclear device at Aldermaston.

The British surveyed the remote Monte Bello Islands under the codename Epicure, the first of many codenames, to ensure that the area would be suitable to test Britain's first ever atomic weapon. The agreement stitched up during that phone call still resonates.

Maralinga was neither Australia's nor Britain's finest hour. Both countries behaved at times with questionable ethics and little regard for future consequences. Later investigations revealed that insufficient safeguards were in place to protect people and land, even allowing for the less developed understanding of matters atomic back then. The harm done to the Indigenous population was substantial and shameful. The test authorities said openly at the time that there was 'nothing to suffer damage except spinifex and mulga' at Maralinga, despite the long and complex history of Indigenous presence there. One top-secret document prepared by the Australian minister for Supply Howard Beale when planning for the permanent test range said, 'Revocation of an existing aborigines' reserve would be involved ... this could be achieved without undue difficulty as the area has not been used by aborigines for some years'. This statement was false.

Most of the events at Maralinga and the other nuclear test sites were top-secret. Today it may come as a surprise to the average person that Australia had a central place in the development of the atomic bomb. School history curricula tend not to mention this fact. Yet, while this country sacrificed much to assist Britain's aspirations to become a nuclear nation, we did not benefit from it. The evidence suggests the opposite. The UK became the world's third atomic power, after the US and the Union of Soviet Socialist Republics (USSR), while Australia was left with a radioactive contamination problem that cost tens of millions of dollars to mitigate. The report of the Royal Commission in the mid-1980s succinctly described Menzies' actions in making Australian territory available without strong safeguards as both 'grovelling' and 'insouciant' – two words that capture perfectly the tone of controlled anger displayed throughout the report. The terms of the agreement struck between Australia and Britain, loosely worded as they were, were not to Australia's advantage in either word or spirit. It is hard to imagine another country accepting the same conditions. Australia accepted them without any particularly strong overt pressure from the UK and even volunteered to bear part of the cost, which the British had not requested. The weight of colonial history provided the true pressure, reflecting how Australia saw itself in relation to Britain at that time.

Canada, suggested in the late 1940s as a possible test location for British bombs, was in many ways a more logical ally in nuclear weapons development. Like Australia, and in contrast to the UK, it had large swathes of lightly populated territory. Unlike Australia, it also had a well-developed research effort in the field and existing collaborations. Canada had a formal nuclear technology development relationship with the US and Britain – the ABC partnership – as part of the Manhattan Project. This gave Canada far higher status than Australia in the world's small nuclear club, a status that would have ensured Canada a greater share of the fruits of the nuclear weapons research had the tests gone ahead there. Indeed, the British dangled the carrot of detailed weapons design information in front of the Canadians. Later, in 1963, Canada even began its own nuclear weapons development program before abandoning it and divesting itself of its permanently stationed nuclear weapons of US origin in 1984.

The UK couldn't have access to the US test sites, so Canada was the next choice. The British surveyed seven sites there and favoured the remote northerly port of Churchill in Hudson Bay, part of the Province of Manitoba. However, when the Canadians learned that the British intended to conduct at least 12 major atomic bomb tests that would severely contaminate a new 450-metre circle each time, they swiftly declined. The Canadians were a little too concerned to protect their own interests.

Australia did not have the same standing in British eyes as Canada. Although both countries were former colonies, Australia had no form at all in the field. Until the postwar era, the best Australian physicists went abroad to do their research, including the great Australian physicist Mark Oliphant, who launched his formidable career at Cambridge's legendary Cavendish Laboratory as a student of nuclear physics pioneer Ernest Rutherford. Australian nuclear physics research really got started when Oliphant, back in Australia, lured Ernest Titterton from the UK in the early 1950s. Titterton set up the Department of Nuclear Physics at Canberra's fledgling Australian National University (ANU). The British atomic weapons test plan was being formulated at the time, and Titterton is prominent in the Maralinga story. The two men fell out though. Oliphant, one of the world's most eminent scientists, was vociferously opposed to scientific secrecy and was considered by the Americans to be a security risk. The test authorities shunned him when he later became a critic of the nuclear tests in Australia.

This story is not as simple as the oppression of a former colony by a fading imperial power, however. Australia entered into the agreement with considerable ambitions of its own. The Menzies government had its reasons, not all of them sycophantic. One incentive was to maximise the value of the country's newly discovered and extensive uranium resources. Uranium was the raw material for both atomic weaponry and atomic energy, but few countries in the world possessed it in such large and accessible quantities. Second, the Australian Government believed that if nuclear war loomed, assisting Britain with its nuclear program would help guarantee Australia's own protection by Britain at least, and possibly the US as well. A third reason was that in the 1950s, Australia toyed with the idea of both civilian nuclear power and its own nuclear weaponry. Who better to learn from than the British (especially as the US would not countenance the idea)? But none of these ulterior motives came to fruition.

This story of many parts is also a Cold War tale. After the end of World War II, the British wartime leader Winston Churchill declared that an 'iron curtain' had descended across Europe. This ideological divide – between the West on one side and the communist nations headed by the USSR on the other – soon sparked an arms race based upon the devastating new weapons demonstrated in Japan. The Soviet Union, with considerable input from the atomic spies who feature later in this book, tested its own atomic bomb just four years later, in 1949.

The Cold War brought secrecy and suspicion into the dealings not just between enemies, but also between allies. In Australia, the Cold War ruptured security relationships with both Britain and the US. A spy ring uncovered after the war at the Soviet Embassy in Canberra implicated a number of Australian public servants (although no charges were laid). The British rocket tests at Woomera, also in the South Australian desert, were temporarily suspended because of these security concerns. Australia was forced to convince both the UK and the US that it could keep security secrets. Australia's domestic spy service, the Australian Security Intelligence Organisation (ASIO), was established in 1949 during the dying days of the Chifley Labor government, under explicit pressure from the two allies. Despite the advent of ASIO, and the even more shadowy Australian Secret Intelligence Service in 1952, neither Britain nor the US really trusted Australia. In the end, Britain provided no nuclear secrets to Australia, and Australia was peculiarly reluctant to ask for them, even when they were being gathered on its own soil.

This is a story of scientific progress as well, and particularly the relatively new science of nuclear physics. Many of the main protagonists in the Maralinga tale were physicists. Some were well inside the Maralinga tent, such as the head of the series, William Penney, and the scientist often said to have been 'planted on Menzies', Ernest Titterton. Titterton was famously characterised as a Dr Strangelove figure, and his reputation was trashed during the McClelland Royal Commission. Penney's reputation came out the other side rather better, though still damaged by the cloak and dagger. Other scientists, particularly the Australians Mark Oliphant and Hedley Marston, were on the outer. They had grave doubts about the nuclear tests in Australia and paid a professional price for raising them.

The science itself is amazing. A once largely worthless heavy element, uranium, had suddenly and dramatically revealed its hidden explosive energy potential at the beginning of World War II. Physicists working in Britain recognised the significance of 'splitting the atom' and developed practical ideas about how to fashion an explosive device. They handed these over to the US Manhattan Project. Within six years, the basic physics that had brought to light hitherto unknown capacities in uranium had resulted in a bomb powered by uranium being dropped on Hiroshima. The bomb dropped on Nagasaki three days later was powered by plutonium, a step-up in technology. Plutonium, a most unnatural and dangerous material, is one of the most important things to understand about Maralinga, because when plutonium fell to earth there it changed the landscape forever.

Australia's media underwent a profound transition during the decades of this story. The articles published in the Australian media at the time of the nuclear tests, and particularly in the early years, were often deferential to Great Britain, overtly patriotic, uncritical of atomic weaponry or actively in favour of it, focused almost exclusively on storylines provided by official information, and lacking scientific detail or analysis. Almost always, statements from test personnel and from the Australian Government immediately allayed any safety concerns raised in these stories. Many of these assurances were shown later to be unfounded. A few contemporary stories were critical of delays to scheduled tests or raised questions about the safety of Indigenous people in the area and the cost-effectiveness of the Maralinga facility. Some were apparently motivated by ideological opposition to the federal government. But the general thrust of most stories and editorials was support of the test series and the nuclear ambitions that underpinned it. The high-profile scientists involved, such as Penney and Titterton, were not subjected to scrutiny.

This began to change in the mid-1970s with a series of stories characterised by a productive scepticism towards the governments involved in the testing, a far higher level of scientific literacy and insight, a diversity of sources and a willingness to confront the government with evidence of untruth and cover-up. With hindsight both the initial phase of secrecy and cover-up and the later uncovering seem inevitable. In fact, the same information controls were in operation in the late 1970s, and the Coalition government of the time, under Malcolm Fraser, was no keener to reveal the truth of Maralinga than the Menzies government before it, albeit for different reasons. But the rising voices of aggrieved military veterans and the advocacy of a small number of politicians such as Tom Uren provided new sources. The markedly different ways the British tests were covered by journalists in the two eras can be explained largely by the approach of the media and the anger of those harmed by the tests, not by changes to the operation of government. The journalists did a much better job in the later era, forcing a lot of the story into the light.

In the saga of nuclear colonialism portrayed in this book, a non-nuclear nation ceded part of its territory to an emerging nuclear nation to test the most destructive weapons ever invented. Australia provided the site, the political backing, many of the running costs of the Maralinga range and some of the logistics and military personnel. But the UK was always in charge. The absence of close contemporary scrutiny of these tests by either the Australian Government or the media allowed the test authorities to conduct experiments of exceptionally high risk and lasting danger. Many hundreds of Indigenous people lost access to their homelands and their traditional ways of life, swept away from the desert test sites like detritus. Military personnel from all the countries involved, but especially those of Britain itself, were exposed to radiation that may have made them ill. The test series included particularly dangerous experiments that left significant radioactive contamination at Maralinga. The nuclear tests were not subjected to the media scrutiny and analysis befitting their importance until many years later. In fact, the British nuclear tests are among the most significant events in Australia's history not subjected to contemporary media scrutiny.

What are we to make of the events at Maralinga in the 1950s and 1960s? Australia was not a nuclear power. The nation was in a highly ambiguous position – it was the staging ground for nuclear weapons testing, but the tests themselves were run with obsessive secrecy and control by another nation, the 'mother country' herself. This made Australia, at least initially, curiously powerless and inept in dealing with the tests. The absence of media coverage and public debate created a gap in most people's understanding of Maralinga, making it in many ways a uniquely tangled national issue, still obscure and perplexing. The fallout from nuclear colonialism in Australia was plutonium-soaked land, certainly, but also growing recognition of the risks inherent in abdicating control over the nation's destiny. The mysteries of Maralinga and its toxic legacy continue to haunt Australia as the red dust of the old desert test site still swirls and the thunder echoes across the plain.

1

Maralinga buried, uncovered

It was a dry wind,
And it swept across the desert
And it curled into the circle of birth
And the dead sand,
Falling on the children
The mothers and the fathers
And the automatic earth

Paul Simon, 'The Boy in the Bubble', *Graceland*, 1986.

Mid-May 1984, autumn, and the plains of Maralinga are cooling down after a hot summer. At the moment it doesn't get much above 22 degrees Celsius during the day, unlike in summer, when the daytime temperatures can exceed the mid-40s. At this time of year it goes down to a chilly 5 degrees Celsius overnight. The Maralinga lands are to the north of the Nullarbor Plain, on the eastern edge of the Great Victoria Desert, Australia's largest desert. Maralinga is 850 kilometres from Adelaide and just north of the Indian–Pacific train line that carries passengers and freight across the continent. Tietkens Well, dug by the English explorer William Henry Tietkens in 1879, is the earliest token of European presence. The landscape is mostly flat, with some gently sloping hills on the horizon and many sand dunes. Most of the terrain is capped by rugged travertine limestone, up to 3 metres thick in places, forming a rocky crust. Its top surface has been busy eroding over millennia into dust that swirls constantly and often whips up into fierce, blinding storms. The overwhelming colour of the landscape is red, broken by the olive green of the stunted, scrubby saltbush, the needle-leaved mulga and the tussocky spinifex that dominate the vegetation. Bird life abounds – there are over 100 species in the area, including bellbirds, honeyeaters, bustards and kingfishers, and bird song is one of the dominant sounds, other than the wind.

The abandoned Maralinga atomic weapons testing range forms part of the western extremes of the much larger Woomera Prohibited Area, a chunk of South Australia that could accommodate England within its boundaries. The British used Woomera for even longer than Maralinga, to test postwar rocket technology. The entire area was surveyed by the legendary Australian bushman Len Beadell, who lent an air of larrikin myth to this vast expanse of outback. He told rollicking tales of his surveying adventures during the mid-1950s, assessing the land for its usefulness for testing atomic weapons. The surveying project was top-secret at the time, though Beadell later wrote about it in two popular books that told the story of his time in the bush. Beadell and his men liked what they saw at X300, as they dubbed the area, and reported back to Professor William Penney, the head of the British nuclear weapons test authority, that it would be perfect for the task. Penney visited to see for himself, spirited there secretly in October 1953, and was well pleased. 'It's the cat's whiskers', he said. History would soon follow.

When it did, this ancient landscape saw some remarkable sights. Hydrogen-filled balloons bobbing around in the bright sky, bristling with measuring gear. Fifty-five-tonne metal scaffolds called, inexplicably, feather beds rising from concrete pads to hold simulated nuclear warheads that glinted briefly in the sun before being blown sky-high with trinitrotoluene (TNT). A Royal Air Force (RAF) Valiant aircraft releasing a bomb 10 000 metres in the sky, creating a mushroom cloud 150 metres above the plain as men in summer-issue military shorts turned their backs to its sun-like brightness.

Fifty-two-tonne Centurion tanks and other military vehicles scattered around seven major bomb sites. A village with a cinema, a swimming pool and single-men's quarters made from prefabricated garages eventually large enough to accommodate thousands of men. An airstrip that felt the weight of both military and civilian aircraft. A network of sealed roads covering 130 kilometres that made travel around the site fast and easy. Then, suddenly, the forces of history departed and the site fell silent. But never again would it be pristine.

In May 1984 virtually no rain will fall at Maralinga – only 22.6 millimetres for the whole month, falling in small bursts on four days. The bright, mild and mostly rainless days are conducive to the meticulous scientific testing of the radiation physics of the site, a search for any remaining traces of the radioactive elements let loose by the British nuclear tests. The scientists are making use of the most fundamental properties of radioactivity to do their search. Over time, radioactive elements such as uranium and plutonium undergo a physical transformation. The unique nature of these elements means that they emit several different kinds of electromagnetic rays (alpha, beta and gamma rays). As they do so, their properties and even their mass change. These rays have fundamentally different properties. Alpha radiation, made up of energetic streams of positively charged particles, is easily thwarted – alpha radiation can't penetrate thick paper. Beta radiation consists of beams of electrons, which have a negative charge. It can penetrate more deeply, but a sheet of light metal such as aluminium will stop it. Gamma rays are like x-rays and can be stopped only by heavy materials such as lead.

These rays and other products of the unique nuclear physics of radioactivity can be detected. The scientists have come equipped to do this – to use their field equipment and later laboratory analysis to work out how much radioactivity this old site actually contains. The scientific authorities charged with this task are confident that they know what is there, based on information provided to Australia by the British. This due diligence survey will simply confirm the past surveys and reports. Dr Geoff Williams, Dr Malcolm Cooper

and Mr Peter Burns are part of the small team of radiation specialists at Maralinga. Their job is to conduct some routine scientific
investigations of the site so it can be officially handed back from
the federal government to South Australia. After that, the land will
be returned at last to its traditional owners, the Maralinga Tjarutja
people, who were displaced and dispersed in their hundreds by the
British nuclear weapons tests that ended over 20 years ago.

At the time of this expedition to the site, which also includes
scientists from the South Australian Health Commission, Williams,
Cooper and Burns work for an organisation called the Australian
Radiation Laboratory (ARL). It was once called the Commonwealth X-Ray and Radium Laboratory and will later change its
name to the Australian Radiation Protection and Nuclear Safety
Agency (ARPANSA) as its role shifts with the imperatives of the
day. Each of these scientists will end up being involved in the story
of Maralinga for years after this eye-opening trip to the South Australian desert. What they will find here will shock them, and soon
after, the many layers of secrecy that have buried Maralinga will be
stripped away.

For Geoff Williams in particular this visit starts a long association. He does not know about Maralinga from his schooling at
Balwyn High in Victoria or his science degrees at Melbourne University; he first heard the name in 1978 as a young postdoctoral
researcher working in the School of Molecular Sciences at Sussex
University in the UK. Before that, like just about everyone growing
up in Australia at the time, Williams knew nothing about Maralinga,
a great Australian secret, barely recognised as part of this nation's
history. This trip is the first of dozens of expeditions he will make
here on his way to becoming a leading expert on radioactive waste
safety. He loves the beauty of the place, the sand dunes and mulga
and birds, and takes the occasional moment to savour it.

By 1984, some aspects of the Maralinga story have been in the
media for a few years. The South Australian media started to take an
interest in 1976. The tone of their stories has been mostly negative,

in contrast to the media reports from when the British nuclear tests were underway. By the 1970s, the events at Maralinga and the other test sites were no longer viewed with the same patriotic equanimity as they were at the time. An Adelaide taxi driver has told Williams that the people of Adelaide once regarded the Maralinga operation as a flag-waving exercise, and they had been proud to help Britain become great again following the war. But now, the driver said, you won't find a person in Adelaide with a good word to say about the nuclear tests.

Despite the newly acquired media scepticism about the British tests, though, the scientific community is reasonably confident that the anecdotes of non-experts who know nothing about radioactivity overstate the dangers of the site. After all, the British scientists and military personnel did surveys and clean-ups before they left, and while a significant portion of this information remains top-secret, those in the know believe the site will be safe for the Indigenous owners to take over. The director of ARL Dr Keith Lokan has taken the prescient decision that the scientists working at Maralinga should not have formal security clearances, so as not to be tainted by reading the classified British record. So the Australian scientists will be able to speak about and publish freely everything that they discover at Maralinga, unconstrained by secrecy laws. Significantly, Maralinga is so far the only former nuclear weapons test site that has reverted from military to civilian hands. Once the handover is complete, everyone can move on. It feels like an obscure era in Australian history that is fast receding in the memories of the few people directly involved.

The ARL team has access to reports left behind by the British nuclear test authorities not classified as 'top-secret atomic' that tell them more or less what to expect. An abridged version of the Pearce Report, compiled by the British nuclear physicist Noah Pearce, sets out an account of the physical conditions at the site.

Pearce was part of the legendary nuclear elite, the small handful of sound inner-circle scientists from the AWRE headed by the dapper and distinguished Professor William Penney, the leader of the British tests in Australia. Pearce came to Australia for the first British test, Hurricane, at the Monte Bello Islands in 1952. He was involved with the two Totem bomb tests at Emu Field in 1953. He did not come to Maralinga until 1958 because he was diverted to the British hydrogen bomb tests in the Pacific at Christmas Island (later Kiritimati), now part of Kiribati. But in the early 1960s, he directed safety arrangements for the highly dangerous plutonium tests from his home base in Aldermaston, visiting Maralinga only occasionally. After overseeing the clean-up operations at the site, he wrote a report. A heavily edited version of this report has been publicly available since 1979, tabled in federal parliament after media pressure, but the full report is secret, open only to those with sufficient security clearance. The federal minister for Mines and Energy Peter Walsh will release the full report later in 1984 as a direct response to the ARL's Maralinga trip.

The Pearce Report has created considerable confusion over the decades because it is totally wrong about some centrally important things, most notably the level of plutonium contamination. More than 22 kilograms of plutonium was exploded in the Vixen B tests at Maralinga. Pearce said that 20 kilograms of this was safely buried in 21 concrete-topped pits dotted around the perimeter of a firing site called Taranaki, about 40 kilometres north of Maralinga village. On the basis of this information, Australia allowed the UK to sign away its responsibilities for the site in 1968. The Atomic Weapons Tests Safety Committee (AWTSC) was the Australian body responsible for monitoring the British testing program to ensure the safety of the Australian environment and its population. John Moroney, its long-time secretary, will later tell the scientists that it would have been 'ungentlemanly' for Australia to question the edicts of the British atomic scientists at the time.

The first week on the ground at Maralinga, which begins on 22 May, proceeds uneventfully. At first, Burns, Cooper, Williams and their team think that the information in the Pearce Report is accurate. They are unable to read it themselves but have been given a summary by Moroney, now head of the Radioactivity Section at ARL. Their initial observations accord with the Pearce assertions – uniform radiation of a microcurie in the old money, or 40 kilobecquerels under the more recent measurement system, per square metre, which is close to normal and no cause for concern. (A kilobecquerel is 1000 becquerels, the international standard unit of radioactivity.) The scientists are not wearing protective clothing; they walk in shorts under the clear Maralinga sun, up and down the dusty grid at Taranaki carrying their radiation gauges. They are careful men, but they know that many myths about radiation have no basis in reality. People ignorant of the physics of radioactivity have a tendency to hysteria – a natural fear of the unseen and unknown. Based upon the data in the Pearce Report the scientists believe that they could stand at the site for hundreds of hours and even, hypothetically, throw handfuls of dust into the air and breathe it in, and still they would get only their annual 'safe' dose.

They start efficiently, knowing that a political circus is about to descend that may slow them down. A delegation headed by the South Australian premier John Bannon and Senator Peter Walsh, accompanied by an entourage of media, is scheduled to arrive on a Royal Australian Air Force (RAAF) plane on 24 May. Although the scientific trip has been arranged for months, it happens to have been scheduled only a month or so after the Australian Broadcasting Corporation's (ABC) investigative public affairs show *Four Corners* screened an exposé about a 'nuclear veteran' dying in Adelaide who blames his service at Maralinga for his illness. By now, Maralinga is a fixture in the Australian media. Veterans have been making allegations since the South Australian RAAF veteran Avon Hudson blew the whistle in 1976 and became the face of nuclear veteran anger; he will continue his campaign for decades, his outrage undiminished by the years. The initial media rumblings of 1976 were followed by a series of landmark investigative reports by Brian Toohey in 1978. Then the Adelaide *Advertiser* ran a high-profile feature series on the plight of the nuclear veterans in 1980. The Maralinga issue, dormant between 1957 (when even superficial media coverage effectively ended, six years before the final atomic experiment) and 1976, is now a media staple.

Radiation scientists can generally do without political and media attention, because it always brings misinformation. Burns, Cooper and Williams have no reason to be glad that Maralinga has bubbled to the surface again. Often journalists approach them for comments about a radiation issue then misquote them in their stories or sensationalise the issue to suit some agenda. Radiation is dangerous, the scientists acknowledge, but the average person thinks it is a thousand times more dangerous than it actually is. The media sometimes play into these fears, distorting radiation research.

If the media have an agenda, so do the politicians. The scientists joke among themselves that the visiting politicians will probably take credit for the scientific expedition, despite it having been arranged long before the current spate of media interest. They are right, of course – an announcement that the scientific expedition is the federal and state government response to the renewed public interest in Maralinga is not far away. The scientists will laugh about it for years to come.

The scientists are not overly concerned with the history of this remote desert location. They are scientists – they are interested only in the evidence they find, not in assigning blame for past political decisions. They come to the task with open eyes as well. They know that the Brits have not been overly forthcoming with assisting in the process of preparing the site for the handover. The British are still being tight with their information about Maralinga, too. A large amount remains classified and unavailable to Australian authorities, years after the site shut down. Nothing has really changed since the days of the tests.

At its heart, this tale turns on a power imbalance between the British test authorities and the country that provided the expansive territory they needed to set up a permanent nuclear test facility.

Changing global attitudes to nuclear weapons and international agreements to limit atmospheric nuclear testing meant that the permanent site was in active service for only seven years. It has lain more or less idle since 1963, other than some clean-up operations.

Maralinga has been of no interest to the British since they struck the 1968 agreement with Australia, ending their responsibilities to the site. The agreement was predicated on the assumption that Britain thoroughly decontaminated the site and cleared the debris to the satisfaction of the Australian Government. In 1977, political pressure forced another survey, carried out by ARL and the Australian Atomic Energy Commission, which perpetuated the conclusions of the Pearce Report. So, too, did the 1983 report AIRAC (Australian Ionising Radiation Advisory Council) 9, which will soon be discredited. None of these reports has come close to the truth. Instead they have clouded the issue both politically and scientifically. In 1979, though, after a major public outcry sparked by Brian Toohey's stories in the *Financial Review* in October 1978, the British were forced to send military personnel to remove drums full of salt imbued with plutonium buried at the Maralinga airfield. They did so reluctantly and only after prolonged negotiations. Pearce himself visited Maralinga for 48 hours in 1978 during those negotiations and reaffirmed his 1968 report. That incident was nothing more than an inconvenience.

The scientists get on with their environmental monitoring work as best they can, diverting briefly for the visit from the pollies and the journos, who drop by for four hours on Thursday. ARL head Keith Lokan arrives with the official party and helps to show them around, explaining what the scientists are doing. The day is mild and dry. The scientists have not detected any serious contamination, but they have found a few areas that show higher than expected activity. Lokan briefs the journalists about these 'hotspots' – small areas of intense radiation – during the visit. A famous moment occurs when he searches for one that was previously discovered. The political party stand outside the fenced-off TM101 testing site, about 20 kilometres to the northeast of the old Maralinga village. Suddenly the radiation monitor starts screaming, right under the foot of one of the male journalists, who quickly jumps back. Peter Walsh's tough female press secretary instantly remarks, 'Now your balls will drop off!' to laughs all round. Among the media is the British journalist Sue Lloyd-Roberts, who will later co-write a book on Maralinga with Denys Blakemore, *Fields of Thunder: Testing Britain's Bomb*. She looks around the abandoned site, with its barbed wire, radiation warning symbols and concrete-topped pits, and writes, 'More ominously, teams from the Australian Radiation Laboratory guiding the ministerial team showed the presence of radioactive material on the surface of the range with their constantly clicking Geiger counters'.

The political and media trip to the site is largely symbolic, designed to distance the two Labor governments – the new Hawke government in Canberra and the Bannon state government – from the Maralinga fallout that dogged the previous Fraser government. As Lloyd-Roberts writes, 'The representatives of the Federal and South Australian Governments were there jointly to express their regret that the atomic test series had ever been allowed to take place in Australia and to pledge their support for all investigations into the possible harm done to servicemen, Aborigines and the environment'.

The Australian newspaper has sent reporter John Stanton, who quotes Walsh and Bannon, and ARL director Lokan, as they survey the contaminated areas. Lokan wields handheld Geiger counters. The radiation chief, while pointing to the dangers of plutonium as a cause of lung cancer, also wishes to allay any fears of the visiting party, or longer term visitors, about the immediate dangers of the area. Stanton quotes him saying, 'There was no evidence the ploughed plutonium was being further spread by wind erosion. But the pollutant could be present in dust, although this would generally have to be breathed for long periods before it would pose a health risk'.

When the politicians and journalists depart, the scientists get back to work. According to a prearranged strategy, they start their survey on the outer limits of the enormous range and work their way in. Part of their work involves field observations using their Geiger counters, and the rest involves measurements of soil samples back at their laboratory in Melbourne. The samples are taken in a standard way using a cylindrical corer device. Three samples are taken from each small area, the soil combined and sieved through a 1-millimetre mesh to remove larger stones and then stored in bags. Later they will be scanned for gamma ray emissions, an excellent diagnostic test for americium-241, which indicates the presence of alpha-emitting plutonium. Because radiation abounds in nature, the scientists also pick up soil samples from outside the weapons test areas to check for naturally occurring uranium, unrelated to the nuclear tests, that may contribute to their readings.

The scientists take in their surroundings. They see lots of small pieces of rusty metal lying around the concrete firing pads, some old pieces of metal tubing or boxes and other scraps. The large infrastructure on the range, though, is long gone – either blown up during the experiments, buried in pits during the 1967 clean-up known as Operation Brumby or removed from the site. The scientists see some electrical components such as cables and connectors, but they find no radioactivity on them. In 1979, several new concrete plinths were erected at Taranaki, and at Emu Field further north, adorned with messages advising visitors to the sites that nuclear tests had taken place there. Before that, back to 1967, visitors would have had little idea of this, since the test era fences and signs were removed in Operation Brumby.

The laborious field measurements and sample collection proceed smoothly. To obtain data on the state of contamination at the site, the Australian team divides the Maralinga range into a grid and walks from end to end taking and recording measurements as they go, as the British Radiation Survey (RADSUR) team did in 1966. They cut the site up into 600 metre by 1 kilometre rectangles and walk up and down, taking a measurement every 20 metres. To preserve the battery power of the field monitors, the instruments are turned off after each measurement and the clicking sound is not activated. This means that the area between each measurement stop is not monitored, a gap that is later recognised as significant.

They slog through this routine for six days without incident and all seems well. The Pearce Report seems more or less to reflect the reality on the ground, with only a few anomalies or hotspots. But on Sunday things start to get strange. If ever the biblical commandment for a day of rest made sense, it is after spending a long, hot, dusty week routinely walking backwards and forwards slowly covering a large swathe of the vast testing site. Most of the field staff take the opportunity to drive north for a day's visit to the Emu Field test site. Burns, Cooper and Williams stay behind to 'have a play around' at Taranaki.

Suddenly, the readings begin revealing multiple radiation hotspots that are definitely unexpected. The Pearce Report suggests the levels of contamination will remain fairly stable, and smoothly continuous – with no big changes over a 100-metre stretch, for example. Yet Burns, Cooper and Williams start to get wildly fluctuating readings over relatively short distances. This is not supposed to be happening. The scientists start to think about putting on face masks and other protective gear. At the Taranaki firing pads they are intrigued to find themselves kicking plutonium-soaked lumps of metal, rock and soil with their boots. Their Geiger counters go berserk near the hotspots. They find radioactive material in many places, particularly around the firing pads. The last of the big 'mushroom cloud' bomb detonations was conducted here. Between 1960 and 1963 it was also the site for the totally secret Vixen B radiological experiments. On 12 occasions during those years, simulated nuclear warheads containing plutonium were blown up using TNT.

No-one can remember now why part of the Maralinga range is named after a region in the North Island of New Zealand. Perhaps it is related to the fact that New Zealand military personnel were part of the uniformed force stationed at Maralinga. Maybe it is an echo of the brilliant New Zealand scientist and Nobel

Laureate Ernest Rutherford, whose basic nuclear physics research at Cambridge University made nuclear weaponry possible (although Rutherford was born on the South Island). Some people have speculated that the name was intended to be Tarakan, site of a battle during World War II in Borneo, but it was written down wrongly and the typo stuck. Certainly some of the other Maralinga test sites, such as Wewak and Biak, reference names familiar to Australian troops who served in Papua New Guinea during the war. Whatever the reason for the name, Taranaki will soon become infamous for its extensive radioactive contamination, the greatest amount on the Maralinga range. In fact, Taranaki will soon be revealed as one of the most contaminated places on the planet. The scientists, like all Australian officials, believed the plutonium was safely buried there, not spread around the landscape. When they report their findings, pressure will grow on the Hawke government, and soon a Royal Commission will be called that will systematically review what happened during the atomic tests. Amazingly, up to that time, the sovereign government of Australia actually has no idea.

The scientists are perplexed by the wildly jumping radiation detectors and review their measurements. 'I had a monitor in my hand, switched the speaker on and monitored the ground', Peter Burns will say later, during an interview with the *New Scientist* journalist Ian Anderson.

> I ended up drawing a circle of about 2 inches in diameter, and you could sort of see the whole of the [radio]activity within that one little area. And Mal [Malcolm Cooper] got a shovel and dug it up and put it in a plastic bag. We monitored the hole again and there was nothing there, so we started squeezing around in this plastic bag and there was a lump of metal in the bottom.

Williams uses two garden trowels bought at his local hardware store in Melbourne to investigate. He halves the soil sample repeatedly, with smaller and smaller portions containing all the activity, until he isolates the small metallic grey lump. Williams holds it in the palm of his hand – a discrete piece of blown-apart bomb debris loaded with plutonium. 'Until that time, we had no idea we were looking for bits and pieces like that', Burns will say.

The British team testing the site in 1966 took soil samples back to the UK, but not the metallic lumps. The soil samples showed nothing untoward back then. Routine soil sampling would be highly unlikely to capture a discrete lump of extremely active material. But the ARL team now finds large quantities of plutonium on the ground.

Plutonium. The material is not found in nature. It is created in a nuclear reactor by bombarding uranium with fundamental atomic particles called neutrons. In the words of nuclear chemist Glenn Seaborg, one of the team who created this dense, silvery substance in February 1941, plutonium 'is unique among all of the chemical elements. And it is fiendishly toxic, even in small amounts'.

Various kinds of plutonium were used in bomb experiments at Maralinga, but one of grave concern is particularly abundant – an isotope of plutonium known as plutonium-239 or ^{239}Pu. Isotopes of an element have the same number of protons but differing numbers of neutrons. Protons define the elements, while neutrons can vary their properties. This means that while it is the same element chemically, each isotope has slightly different nuclear properties, and this leads to different physical behaviour. Plutonium in fact has up to 20 isotopes, each with the same number of protons but different numbers of neutrons. The highly toxic plutonium-239 has 145 neutrons and 94 protons in its nucleus, giving it a total atomic weight of 239.

All the plutonium isotopes are radioactive and undergo a process of 'decay', releasing radioactivity in several different forms and eventually turning into other elements over time. The radioactive half-life (the time required for half of the nuclei of a radioactive isotope to undergo radioactive decay) of plutonium-239 is more than 24 000 years, much longer than the other plutonium isotopes. This long half-life means that plutonium-239 will be present in the environment so far into the future that it might as well be called forever.

The persistence of its radioactivity is not the only reason plutonium-239 is especially dangerous. Even in small doses, it can cause terrible damage if absorbed into the body. By the time of the 1984 expedition the scientists know that it is subject to the strictest of controls; when the British were at Maralinga it was released onto the open range. To make it worse, much of the Maralinga plutonium was turned by the Vixen B tests into a fine form that could be inhaled. This made it potentially hazardous for anyone who encountered the dust of the area. The risks are well known: if you inhale 20 milligrams you will probably die of pulmonary fibrosis within a month. Inhaling a milligram will lead to lung cancer. The strange thing about plutonium is that it is relatively safe outside the body, and in fact you can touch a lump of it (as Williams does that Sunday). The alpha rays that it emits would not get past your skin. But once it's inside the body, it turns deadly. It can be inhaled into the lungs, ingested through the mouth or absorbed through a wound, and if it enters the body through these pathways, there is a strong statistical probability that it will cause various kinds of cancers.

The Pearce Report does not mention any fragments contaminated by plutonium. As a consequence, the AWTSC, whose job was to oversee safety standards at the Maralinga range during the tests and afterwards, disregarded the possibility that later visitors to the site might unknowingly pick up these fragments and take them away.

The fragments are ejecta, metallic debris from firing simulated nuclear warheads during the Vixen B experiments. These so-called minor tests left terrible contamination, far greater than the more dramatic mushroom clouds. The simulated warheads containing plutonium were exploded using TNT. As a direct result the metallic scaffolds that held the assemblies aloft, the oddly named feather beds, were imbued with plutonium-239, as the ARL scientists, to their growing concern, are now discovering. The nature of the fragments varies. In some places, they will say in the report to be written directly after this landmark survey, are fractured pieces of steel, light alloy or other material coated with plutonium. The most radio-active piece found is a concave trapezoidal sheet of 12-millimetre steel plate, about 250 millimetres long and 120 millimetres wide, roughly the dimensions of a piece of A4 paper folded lengthways and pulled slightly out of shape. This metal has a massive 7 gigabec-querels (7 billion becquerels) of plutonium-239 on its inner surface. Most of the fragments are smaller, though, ranging from about half a millimetre to a few centimetres in length.

The scientists also find evidence of plumes. These elongated hand-shapes on the ground trace the curves of the great clouds of fine plutonium oxide particles that lifted 1000 metres with each Vixen B detonation, were dispersed in the direction of the prevail-ing winds, then drifted down to the surface. The plumes splay out to the west, northwest, north and northeast of the firing pads. They can be detected because the plutonium carried back down still sits close to the surface.

The Pearce Report says that 20 kilograms of the 22.2 kilograms of plutonium-239 is safely buried in pits at Taranaki, bulldozed and sealed in years ago. Why are the ARL scientists finding lumps of the stuff, and plumes picked out in surface-dwelling plutonium? An evenly dispersed sprinkling of sparse tiny particles would barely trouble the Geiger counters. But they are finding a web of hotspots that together contain kilograms of the most deadly form of pluto-nium known. It turns out the Pearce team back in the 1960s was made up of low-rank military personnel told to take various meas-urements without ever understanding the physics of what they were doing. Their measurements were effectively worthless. Because of the spotty nature of the plutonium hotspots, their monitoring methods totally missed the plutonium scattered all over the Maralinga range. The Australian Government accepted the return of the site from the British in 1968 on the basis of a fundamentally flawed report.

As Peter Burns will drily observe years later, 'If they had been as far out in their design of the bomb, they would never have been allowed to build atom bombs in the first place'.

The analytical techniques of 1984 are more sophisticated than those of the 1960s, but there was still enough knowledge to survey the area properly back then. The wonder is why the British chose not to do the job they should have done but instead ordered untrained junior troops to walk around a relatively small part of the range taking surface alpha radiation measurements and picking up a few soil samples for analysis in the UK. Alpha radiation measurements are notoriously difficult to detect in the field because alpha particles emitted by radioactive elements travel only a short distance through the air. 'Take the lackeys out there, start at ground zero, take a compass bearing, walk every 100 metres and measure, which is what they did. And frankly if you go out and try to measure like that now you would get the wrong answer', Burns will later say.

This inadequate technique produced a politically acceptable and expedient outcome but did not get anywhere near the truth. The ARL scientists are using a range of methods, including specially designed portable field probes for detecting the gamma ray emissions from the radioactive element americium, an excellent marker for plutonium in the environment. The peculiar physics of plutonium means that as it emits radioactivity over time, part of the plutonium is transformed into americium in a predictable way. Measuring the ratio of plutonium to americium gives a sensitive gauge of the true plutonium concentrations in the field. By 1966–67, americium would have been present in the surface soil at the site. Although the British did not have the same techniques for measuring americium then, they could have performed other kinds of experiments that would have given them a workable ratio. As Williams will tell Ian Anderson in 1993, 'It is hard to see why they didn't appreciate the physical deficiencies in the method that they used'.

> If an important part of the tests was getting information
> on where the plutonium ended up, the environmental
> consequences, then you would think they would put the effort
> into more scientific thought into getting those measurements

more or less right. We know full well that other countries performing these experiments at the same time certainly did get it right. The Americans got it right. So you would think that the British would have given thought to getting those measurements.

Why did they not put in the effort to get it right? We can only surmise now, but it is at least possible that the British had moved on by 1967 and their focus was to divest themselves of the inhospitably hot and remote Maralinga range. Perhaps they wanted to tie up the loose ends as efficiently and quickly as possible. A superficial and inadequate survey, unsupervised by Australian authorities, produced a convenient outcome. It was good enough.

The British attitude since Australia started becoming restive over the issue in the late 1970s has been an irreconcilable combination of 'there is no risk at the site' and 'if there is a risk, the Australian Government knew about it and accepted it, and it is their problem not ours'. Margaret Thatcher's UK Conservative government will, later in 1984, present a submission to the Royal Commission that will say in part:

The Australian government knew that the nuclear tests to be carried out [at Maralinga] would cause residual contamination and that, for that reason, public access to it would need to be restricted for the foreseeable future … Scientific knowledge is not now, and certainly was not then, sufficiently advanced to enable a complete decontamination of an area in which nuclear explosive tests have taken place. In 1955 [when the agreement to establish Maralinga was being negotiated] the Australian government did not seek such an onerous, if not impossible, undertaking from the UK government, nor would the UK government have committed itself to the use of the Maralinga range if it had contemplated any such requirement.

But, as we will see, the British authorities did not tell the Australians that plutonium-239 would be dispersed in this way at the site. At the time of the agreement to establish Maralinga, the 1960s minor trials were not part of the negotiations. The Australians didn't know exactly what went on at Taranaki until the ARL scientists discovered the reality and the science started pouring in after their landmark survey. Australia should have known, but it didn't.

Soon after the political delegation returns to Canberra from Maralinga, ARL makes contact with one of the visitors, the federal minister responsible for Maralinga, Senator Peter Walsh. They tell him that the site does not conform to the information in the Pearce Report. Walsh immediately announces in parliament that the scientists have found 28 plutonium-contaminated fragments at the site. The ARL scientists have provided the number 28 deliberately at this stage – a number they can verify based on the data available. When they started to get hotspot readings during their survey, the scientists conducted a scan across one of their rectangles, focusing their attention on a narrow area, and documented 28 fragments in that designated zone. Later surveys will ultimately find about three million fragments spread over square kilometres.

The discovery of fragments is highly significant, not just because their presence undermines the assurances given by the British. Before this, everyone connected to the site believed that the major danger would be from inhaling the dust containing small particles of plutonium, although they think that this particular risk is minimal (a view that will soon also be challenged). The fragments suggest new risks. They are highly radioactive and if handled by people – for instance, visitors collecting souvenirs from the site – the radioactivity might enter the body by other means, such as through wounds. The Australians at the site have started to realise what they are facing.

Each fragment is found to be significantly radioactive, measuring about 100 kilobecquerels or more. To put this into perspective, in Australian universities in the 21st century a researcher who wants to do an experiment using radioactive material that is 400 becquerels or more will need a special licence and training, and extensive special handling equipment. This for material that has only one-twenty-fifth the amount of radioactivity of each one of the fragments found at the Maralinga site during May 1984. Moreover, plutonium in these quantities is a safeguardable nuclear material. Under the terms of the Non-Proliferation Treaty of 1968 (which will be extended indefinitely in 1995), to which Australia is a signatory, all material that could be used to create a nuclear weapon has to be declared and prevented from being used for weaponry. Such material should not be lying around on the surface in collectible quantities. Then there are the fingerprints of British atomic weapons fuel remaining in that plutonium, which will shortly be discovered by the ARL scientists in their forensic 'nuclear archaeology' laboratory analyses of the material. The exact way that each nuclear nation makes its bombs is highly classified, yet the material left behind on the ground at Maralinga was open to analysis, thus revealing some of the secrets of the British bomb. Nothing could be more foolhardy or irresponsible, not to mention the mockery made of all the paranoid secrecy during the test series in Australia. Millions of these fragments will eventually be found lying on or just below the surface at Maralinga, readily accessible to anybody casually visiting the area. The Pearce Report contains no hint of this possibility.

A long period of analysis of the Maralinga site will follow this fateful ARL visit. The ARL scientists will document the major discrepancies between the levels of contamination claimed in the Pearce Report and what they have found on the ground and will present several influential reports to the Royal Commission. Later still, even more damning information will detail the magnitude of these discrepancies and British culpability (see chapter 11).

Later in 1984, Peter Walsh speaks several times in federal parliament about Maralinga. This issue is both a growing priority and an irritant to him. The treasurer Paul Keating nicknames him Sid Vicious because of his dour, unsentimental personality and his tough, pragmatic approach to all issues in his portfolio, including Maralinga. In a lengthy statement to parliament on 4 May 1984, Walsh confirmed that he was seeking to release publicly the entire Pearce Report. That week investigative journalist Brian Toohey published his brilliant *National Times* story based in part on the leaked report. In his parliamentary statement, Walsh said, 'Let me assure the Senate and the Australian people that this Government has no interest or intention of keeping facts relating to the nuclear tests in Australia secret'.

In June 1984, Walsh addresses parliament after receiving a chronology of the British nuclear tests that he commissioned from the physicist John Symonds, a consultant to his department. Symonds will later prepare an exhaustive account of the British tests for the Royal Commission. In this parliamentary statement, Walsh refers to the particularly problematic minor trials:

> It is clear that it is these trials, and particularly the Vixen B series, which involved the use of plutonium, that produced the major source of radiological contamination which remains of concern at Maralinga today. One would assume that the Australian governments of the day were aware of the nature of these tests. However, Australian documents examined to date do not enable us to determine this.

This hints at the revelations to come in the Royal Commission, when Justice James McClelland – Diamond Jim – will begin the process of making public just how little the Australian Government actually did know at the time. Walsh announces the Royal Commission into the British nuclear tests on 5 July 1984. In his media release, he says the inquiry is charged in particular with examining 'measures that were taken for protection of persons against the harmful effects of ionising radiation and the dispersal of radioactive substances and toxic materials as judged against standards applicable at the time and with reference to standards of today'.

Investigative journalists have begun sniffing around, including Howard Conkey and Paul Malone from the *Canberra Times*. Conkey writes to Walsh in June 1984 seeking clarification on the contaminated material found by the ARL team. Walsh replies on 28 June, 'The present concerns with the newly identified fragments arise from the recognition that this material could be moved from its present site in an uncontrolled way (for example, picked up in the tyre tread of a motor vehicle)'.

At Maralinga, in that mild desert May of 1984, the scientists have been amazed and shocked at their findings. Over time they will become a little angry as well, particularly when their senior colleague John Moroney crunches the Roller Coaster data in the early 1990s. Williams, Burns and Moroney will be sources for Ian Anderson's landmark story in *New Scientist* in 1993 on the British deceit about contamination at Maralinga. His role in the Anderson story will be Moroney's last act in relation to Maralinga – within days of the story he will die from multiple myeloma. In the end, science, journalism, politics and the anger of veterans will all play a role in uncovering the shadowy Cold War story of the Maralinga desert nuclear test range. To find out how we got to this point, we need to go back to the birth of atomic weaponry.

2

Britain's stealthy march towards the bomb

I'd learned by the bitter path that to touch the pitch of secrecy was to be contaminated for a very long time, that governments and politicians wanted not men who believed in the integrity of natural knowledge but men who would tell them what they want to hear, and that the truth has no meaning for a Churchill ... [or] a Menzies, if it is politically inconvenient.

Professor Marcus (Mark) Oliphant, Australian physicist, 1956.

Secrecy was not only a guard against enemies but a barrier between allies. It caused much wartime ill-will between Britain and the United States.

Margaret Gowing, official historian of the British nuclear energy and weapons development programs, 1978.

'Now I am become death, destroyer of worlds.'

J Robert Oppenheimer, physicist and leader of the Manhattan Project, quoting from Hindu scripture, as the first atomic weapon was tested at Alamogordo, New Mexico, 1945.

Britain came to the red sand of Maralinga through a series of contingent events. These ranged over politics and geography, colonial legacies and scientific pragmatism. One of the most significant factors was Britain's early dominance in nuclear weapons development. Britain incubated the atomic bomb. Although she quickly lost custody to the US, Britain's strong and distinguished tradition of physics research was the foundation for the terrifying weapons that shook the world and reshaped geopolitics after World War II.

British scientific history lays claim to some of the greatest minds in the field, including the father of modern physics, Sir Isaac Newton. At prolific laboratories, such as the Cavendish Laboratory at Cambridge University and the Department of Physics at the University of Birmingham, modern nuclear physics had some of its greatest advances. This is, after all, a story about physics as much as anything else, because physics made nuclear weaponry possible. Many physicists were later conflicted by this, first when atomic weapons killed tens of thousands of Japanese civilians and then when the weapons grew in strength and number until they could destroy our planet many times over.

British physicists were involved at all levels of the push to build the bomb. The country also welcomed and protected many Jewish physicists escaping persecution in Nazi Germany. These refugees found their professional contributions and lives were valued, working with colleagues on this most secret and challenging of problems. Physicists working in Britain drew together the disparate strands from the burgeoning field of nuclear physics – particularly essential research conducted in Paris just before the war – and came up with an ominous synthesis. In short, British physics initially powered the US Manhattan Project, the staggering scientific and technological effort that took just six years to develop the atomic bomb that was loaded into the bomb-bay of an operational aircraft in August 1945.

To understand the British tests in Australia we need to understand the British role in the Manhattan Project. The project drew upon substantial American expertise and technical wizardry, but the British provided the early intellectual boost. Tube Alloys, the world's first atomic weapon development program, exemplified the British approach to developing the bomb. It buried a world-changing scientific research project under an obfuscating and boring bureaucratic name. The Tube Alloys Directorate was established in September 1941, and it continued after the war in several different guises. The secret research and development organisation with its deliberately misleading title quietly worked in a series of nondescript university laboratories and, later, dingy industrial workshops. No big, expensive infrastructure was involved. The can-do and well-funded Americans made the first operational bomb on the back of some remarkable work done on a shoestring in Britain under the banner of Tube Alloys and the Maud Committee that predated it. Debate still rages over whether this was a good thing or not, but either way it happened. The rapid-fire research of the Tube Alloys Directorate was transferred to the US, where it continued on to a dramatic conclusion under a far more recognisable name, the Manhattan Project.

The Manhattan Project changed the world. On 6 August 1945, this secret wartime project came to fruition in the most shocking way when the first atomic bomb was dropped on Hiroshima, followed, three days later, by a different kind of atomic bomb on Nagasaki. Exact figures are disputed, but at least 185000 people died, either immediately or in the months directly after. Single bombs have never created so many casualties. The physics that enabled this broad-scale killing emerged at the beginning of the war; the atomic bombs played a role in ending it.

At the beginning of World War II the US president Franklin Roosevelt received an exceptionally important letter signed by Albert Einstein, the world's best known and possibly greatest physicist. Leo Szilard and Eugene Wigner, both European refugee physicists based in the US, had drafted this epochal missive. The letter urged the president to investigate a totally new kind of bomb, a weapon that would exploit the process of nuclear fission, later known to many as splitting the atom. The atom to be split was from the remarkable element uranium.

This idea had gained its theoretical underpinnings in the late 1930s, so the science was new and almost entirely untested. Like a nuclear chain reaction, though, knowledge began exploding out of the initial idea. As noted by Margaret Gowing, the great historian of British atomic science, between January and June of 1939 over 20 papers were published on uranium in the prominent journal *Nature* alone. Military secrecy soon closed down publication as the work was recognised as strategically valuable and dangerous. But, for a while, an army of nuclear physicists, who saw what power might be unlocked by breaking the forces holding a uranium atom together, joined the frenzy to share their results.

Uranium, nature's 'heaviest' element (that is, with the most matter in its nucleus), had virtually no practical use before 1938 when the Berlin-based German scientists Otto Hahn and Fritz Strassmann formulated a theory of nuclear fission that exploited its unique properties. Hahn, a chemist, built upon theoretical work he had done with the Austrian physicist Lise Meitner the year before and also on research in Paris that established basic understanding of how a uranium atom could be split and what happened when it was. Uranium was already known to bristle with many more neutrons than protons, but the Paris group showed that uranium atoms fling out some of the extra neutrons when they are 'split'. At first the Hahn–Strassmann theory held only intellectual interest, because no-one thought that the energy produced by cleaving an atom could be harnessed and controlled. Nevertheless, the theory led quickly to related ideas of chain reaction and critical mass, explained below, that would, in a practical sense, make a bomb possible. Niels Bohr and John Wheeler published a memorandum on the fission process just three days before the outbreak of the war in Europe that gave scientific impetus to the notion of a nuclear weapon. This, according to Gowing, effectively 'put all belligerents at the same theoretical starting point in the pursuit of an atomic bomb'.

Soon, as the almost unimaginable possibilities presented themselves, both Allied and Axis governments started to crack down on the free publication of this growing body of scholarly research. Some scientists, notably Szilard, believed that publication should be stopped voluntarily, and he encouraged other scientists to refrain from publication even before government-imposed secrecy took over. In the first few months of the war in Europe, as uranium fission research was taken over by government entities and as scientists recognised the gravity of these many new insights, publication ceased altogether. This alerted the Soviet Union to the fact that the science had shifted to the military, and they deployed spies to crack the secret knowledge. The spy rings were forming even before the scientists knew how far their research would take them, and they proved to be devastatingly effective once the bomb moved from theory to reality.

The science can be simply stated. An atom has a nucleus consisting of protons, which have a positive charge, and neutrons, which have no charge. Electrons, which have a negative charge, surround this nucleus. Strong forces hold the nucleus together – forces far stronger than those that hold the electrons around it. The number of protons in its nucleus determines the atomic number of each element, its position on the Periodic Table and its chemical properties. However, the nuclear properties of the atom vary depending on the number of protons *and* neutrons. The key to understanding how nuclear bombs work is the neutron. For heavy atoms with high atomic numbers, the fact that protons have a positive charge becomes significant, because objects with the same electrical charge repel each other. A large nucleus, such as that of uranium, is full of positively charged particles all pushing each other away. Extra neutrons are needed to stabilise it. Therefore, heavy elements tend to have more neutrons than protons in their nuclei. Some heavy elements are known as fissionable, which means their nuclei can be split. Of the fissionable elements, a small number are known as fissile, because they can readily sustain a nuclear chain reaction.

Natural uranium-238, the most abundant isotope, is fissionable, but the rarer uranium-235 is fissile.

Uranium-235 was the essential isotope for the early development of the atomic bomb. If you bombard uranium with neutrons, when the beam of neutrons hits the nucleus, it reacts. As the neutron enters the uranium nucleus at great speed, it splits the nucleus apart violently into two roughly equal fragments, hence the popular phrase splitting the atom. Each fragment has approximately half the mass and half the nuclear charge of the original substance, and the process releases a staggering amount of energy. The pioneer of fission, Otto Frisch, along with his aunt Lise Meitner, calculated it as 200 million electron volts, evenly divided between the two new fragments. These fragments rapidly fly away from each other, carrying their energy with them.

Uranium-235, as a rare fissile element, has an odd number of neutrons (143), which means that for complex physical reasons it can more easily sustain a chain reaction than uranium-238. In effect, it is almost impossible to induce an explosive chain reaction in uranium-238, but relatively straightforward in uranium-235. The neutrons released by uranium-235 following fission can be catapulted into another uranium-235 nucleus, causing a cascade of fissions as atoms are split by the neutrons flung out from a preceding reaction. Each successive round of fissions is stronger than the one that came before. This is a chain reaction. As Gowing put it, 'The reaction will spread from atom to atom through the mass of fissile material. Each fission of a uranium atom results in the release of over a million times as much energy as in the combustion of a carbon atom. A pound of uranium, therefore, if completely fissioned, would yield as much energy as several million pounds of coal'.

Critical mass refers to the fact that a certain threshold amount of fissile material is needed to sustain a chain reaction. More material is needed than will actually react, because not all neutrons that split out of the first reaction actually go on to split other nuclei. Some just depart from the reaction without further interaction. Material reaches critical mass when more neutrons are able to go off and split other nuclei than are lost in the reaction. In the early atomic weapons, just a few pounds of uranium-235 were needed to reach critical mass. Physicists began to grasp these insights into fission, critical mass and chain reaction only in the 1930s, but they came together at exactly the moment in history that they could, for better or worse, be acted upon.

When Einstein, Szilard and Wigner sent their doomsday letter to Roosevelt in August 1939, just days before the war began in Europe, they didn't tell him they were proposing an unsubstantiated idea. Most physicists at that point did not actually believe such a weapon was possible, largely because only a tiny fraction of natural uranium is actually fissile – which meant it would take tonnes of the stuff to have enough material to start a nuclear chain reaction. The letter immediately piqued the president's interest not least because it was signed by Einstein. In October 1939 Roosevelt ordered the creation of the Advisory Committee on Uranium, and the American atomic bomb effort got its tentative start.

By then, more important developments were afoot in the UK, unknown at first to Roosevelt and his newly created uranium committee. Physicists Otto Frisch and Rudolf Peierls concluded that the physics problems associated with achieving critical mass were not insurmountable, if the isotope uranium-235 could be separated from natural uranium-238. The physicists, both Jewish refugees (Frisch was Austrian, Peierls German), were working at the University of Birmingham in England. They deduced that uranium-235 would achieve fast critical mass with only a pound or two of the material, effectively resolving the technological stumbling block to a nuclear weapon. They wrote a memorandum outlining their findings, the famous Frisch–Peierls Memorandum. The canny Australian physicist Marcus Oliphant, usually known as Mark (later Sir Mark), working at the same university, made sure that the right people at the heart of the British Government came to know immediately what was in that brief (three-page) document. This world-changing memorandum began modestly:

The possible construction of 'super-bombs' based on a nuclear chain reaction in uranium has been discussed a great deal and arguments have been brought forward which seemed to exclude this possibility. We wish here to point out and discuss a possibility which seems to have been overlooked in previous discussions.

Frisch and Peierls then made a simple and eloquent case for a bomb made from a small quantity of uranium-235. The device they described would have two separate lumps of the substance that would be rammed together when detonated to instigate a chain reaction. Their words reveal a moment in science when educated insight precipitates a leap forwards in human knowledge. They had no applied research to back up what they were saying. They had their reasoning ability combined with their knowledge base. It was enough.

The scientists also speculated on what might happen if such a bomb were exploded in the real world:

> Any estimates of the effects of this radiation on human beings must be rather uncertain because it is difficult to tell what will happen to the radioactive material after the explosion. Most of it will probably be blown into the air and carried away by the wind. This cloud of radioactive material will kill everybody within a strip estimated to be several miles long. If it rained the danger would be even worse because active material would be carried down to the ground and stick to it, and persons entering the contaminated area would be subjected to dangerous radiations even after days. If 1% of the active material sticks to the debris in the vicinity of the explosion and if the debris is spread over an area of, say, a square mile, any person entering this area would be in serious danger, even several days after the explosion.

In calm language, the scientists were explaining a bomb never before seen. This nightmarish weapon would not just have a huge explosive impact but would go on harming people long after its detonation, and well beyond ground zero. Such a bomb would change the course of any conflict. If these scientists hounded from Nazi Germany could envisage such a weapon, so could the dozen or more highly capable physicists who were still there, all of whom could read the recent outpouring of scientific literature and follow their own intellectual intuitions. This fact haunted the Allies throughout the war, and with good reason. Despite efforts by the British to buy up stocks of available uranium oxide when the possibilities of this previously useless material became apparent, the Germans succeeded in getting their hands on significant quantities. That could mean only one thing.

There was no time to delay. On the basis of the startling new knowledge unleashed by the Frisch–Peierls Memorandum, a British wartime committee of scientists known as the Military Application of Uranium Detonation (MAUD) Committee was established in April 1940 out of the broader but short-lived committee called the Scientific Survey of Air Warfare. (It quickly became the Maud Committee and there are several theories why the upper case acronym was dropped, one being in reference to Maud Ray, a governess to the children of the pioneering quantum mechanics theorist Niels Bohr while he visited the UK.) The Maud Committee developed the ideas sketched out in the Frisch–Peierls Memorandum into a practical plan to build an atomic bomb. The committee had as members some of the leaders of British physics at that time, including Mark Oliphant and the later Nobel Laureate John Cockcroft. The authors of the famous memorandum were not allowed to be members, though, because of their German background and continuing connections with some German physicists. These connections were, most notably, via Frisch's eminent aunt, Austrian-born Lise Meitner, who maintained links to colleagues in Germany even though she had fled to Sweden. Nevertheless, the pair was consulted regularly by the committee. Because so many of the most important thinkers in this field were from Germany, over time the Maud Committee accepted several exiled Germans as members, although never Frisch or Peierls. The pair did become members of a technical sub-committee later, though, where their input continued.

The Maud Committee, which met at the Royal Society in London and was chaired by Professor George Thomson, oversaw a secret experimental program carried out at four main British universities: Liverpool, Birmingham, Oxford and Cambridge. Physicists from the eminent London universities had been evacuated to Liverpool during the Blitz. In an astonishingly short time and for negligible expense, this highly targeted team made huge progress. They quickly filled in the gaps of the theoretical musings of Frisch and Peierls around concentrating uranium-235 with new understanding and practical strategies. According to Gowing, despite the fact that this work was carried out in open laboratories at the universities, the physicists managed to keep what they were doing secret, largely through an ingenious system of codenames, including tube alloys, which soon took on considerable significance. While the exact reasons why many of the British atomic codenames were chosen are usually unclear, the pattern does seem to be obfuscation, often combined with British whimsy.

In these early, pioneering days secrecy was not difficult to achieve. By the end of 1940, as Nazi bombs rained down on Britain during the Blitz, the Maud Committee delivered its verdict – uranium-235 could be created in the right quantities and therefore a fission 'super-bomb' was feasible. Experimental and theoretical hints emerged of new, unnatural radioactive elements too, created in a reactor by irradiating uranium with neutrons. These new theorised substances were named to conform to the planetary origins of the name uranium. One was neptunium. The other was plutonium.

To create a nuclear bomb, the physical properties of the radioactive substance at its heart have to be awakened in a precise and technically challenging way. A fission reaction occurs when a substance such as uranium-235 is bombarded with a controlled cascade of neutrons, unleashing tremendous energy. However, this reaction is fraught with difficulty since bombardment on its own is insufficient. Generally neutrons fired at uranium arrive too fast and are dispersed without causing fission reactions. The neutrons have to be slowed down and better directed to ensure a chain reaction. Adding hydrogen as a moderator, generally in the form of so-called heavy water, can slow neutrons.

Heavy water is made up of hydrogen and oxygen, like normal water, but the hydrogen is a particular isotope called deuterium that contains neutrons. The neutrons make this kind of water about 11 per cent denser than normal water. The particular properties of heavy water enable it to slow down incoming neutrons. By a combination of luck and good management, 26 cans of heavy water produced in France as the war began were spirited away to the UK when Paris fell. Initially stored at Wormwood Scrubs prison, and later shifted to Windsor Castle, the heavy water proved to be an invaluable resource as scientists attempted to understand how to moderate the influx of neutrons. Slowing down the reactions in uranium posed significant problems, and the first bombs could not have been developed so quickly without heavy water. As it happens, plutonium is ruthlessly efficient in creating a chain reaction using fast neutrons, thus removing the technical constraints on slowing neutrons down. Ultimately, plutonium became the material of choice in nuclear bombs, including those tested in Australia.

By March 1941, the Maud Committee had incontrovertible proof that the bomb was feasible and could be created from just 8 kilograms of uranium-235. They finalised their report in July 1941. Bluntly titled 'Use of Uranium for a Bomb', it laid out a blueprint for a workable bomb. The committee also produced a report on atomic energy, 'Use of Uranium as a Source of Power'. The two new uses of uranium were laid bare, and the committee never met again.

At the time, Britain was in the middle of the Blitz, which lasted from September 1940 until May 1941, so it was not a safe place to develop the bomb. They needed to find somewhere else.

The Maud Report on the atomic bomb was sent up the chain of command to the highest levels of the British Government. Several months later a new entity was created: the Directorate of Tube Alloys, housed within the Department of Scientific and Industrial Research. Tube Alloys was Britain's new A-bomb establishment, the forerunner to the organisation that later tested bombs in Australia. The name was marvellously enigmatic – indeed, as Gowing described it, 'meaningless and unintelligible' – and most likely to be associated in the casual observer's mind with aeroplane or tank parts. Such a simple strategy was remarkably successful and provided excellent cover.

While the Maud Committee stage-managed the realisation of the dangerous Frisch–Peierls idea, the US was working on its own top-secret physics separately. The Americans had not yet been party to the Frisch–Peierls Memorandum; if they had, their work might by then have been further advanced than the British. But at this point, Britain was well in front. For the idea of the bomb to be made real, though, it would have to shift across the Atlantic.

Some insiders in Britain wanted the bomb-building infrastructure to be established in Canada, a close and trusted member of the Commonwealth and the main supplier at that stage of uranium oxide, but in the end the better equipped US had to be the choice. After a slow start, American atomic bomb research was accelerating. A summary of the activities of the Maud Committee was transmitted to the US in July 1941, although the Maud Report itself did not arrive until October. News about the Maud scientists and the contents of the Frisch–Peierls Memorandum (so far unseen in the US) had an instant effect. Mark Oliphant was among the insiders who took this knowledge to America in 1941, speaking to US military officials and fellow physicists. He found an attentive reception. Such visits led to a chain reaction of information, and the physics world in America lit up. The American physicists, now in possession of the work by the Maud scientists as well as their own escalating research, understood immediately what could be unleashed.

In December 1941, just after the Japanese attack on Pearl Harbor brought the US into the war, the US Government established the Office of Scientific Research and Development to pursue atomic weapons science. This organisation began collaborating with the new Directorate of Tube Alloys, intensifying the information flow across the Atlantic for a short while. In the US, Glenn Seaborg pioneered research on the chemistry of plutonium at Berkeley in California that proved essential to bomb development. In the Metallurgical Laboratories in Chicago sufficient quantities of plutonium were created for the first time. The Americans, in remarkably quick time, did not actually need the British any more, and increasingly UK physicists were shut out. Political tussles erupted between the two allies over who would take the lead.

By early 1942 there was no doubt. The US was unstoppable. After considerable wrangling, Roosevelt and Britain's prime minister Winston Churchill struck the Quebec Agreement, a painstaking negotiation that set the ground rules for engagement between the two countries and established a joint project to find and buy uranium. The rules limited British access to the project, which meant that British scientists developed expertise only in certain areas. The Americans would not agree to continuing British involvement on any other terms. The Quebec Agreement allowed British scientists to travel to the US to continue fulfilling the promise of the Frisch–Peierls Memorandum, and atomic bomb building activities in the UK were effectively closed down. All efforts were focused on America. The full implications of the Quebec Agreement were felt later when the US pushed them away completely and the British found themselves only partially equipped to build their own atomic weapon.

The Manhattan Project had its beginnings in 1939 when Roosevelt established the Advisory Committee on Uranium. It became a military and political priority in August 1942 when it was transferred to the control of the US Army. The project was named after

New York's Manhattan Island, where a group of engineers had been recruited to construct some of the required infrastructure. A new top-secret laboratory, built on a desert mesa at remote Los Alamos in New Mexico, was set up to build the bomb. Many British scientists went there to brave the desert winds and fight against time to build a weapon never seen before. Other laboratories around the country, notably the Lawrence Livermore National Laboratory in California and Oak Ridge, built on a farm in Tennessee, joined the effort as well. In all, around half a million people ended up working on the Manhattan Project.

From 1942, General Leslie Groves, a gung-ho US Army officer, led the atomic bomb project. Groves chose a young American physicist, J Robert Oppenheimer, to head the weapons laboratory. It was a surprising choice because Oppenheimer was open about his leftist leanings – some even thought he was a communist – and throughout the entire time he worked on the project he was enthusiastically investigated by the notoriously paranoid Federal Bureau of Investigation headed by J Edgar Hoover.

Groves and Oppenheimer gathered together a team of physics brainpower the likes of which had never worked together before. Hundreds of thousands of physicists, chemists and technicians joined the effort between 1941 and 1945, including some of the greatest contemporary thinkers. Many were European scientists who had fled the rise of Nazism. Once the political differences between America and Britain were sorted out, largely through the 1943 Quebec Agreement, a significant number of scientists joined the Manhattan Project from the UK as part of the British mission, including William Penney and Ernest Titterton, later pivotal in the Maralinga story. Klaus Fuchs – another pivotal scientist for a different reason – also joined the effort.

Physics had always been an open science, where an international community of theorists and experimentalists shared their hypotheses and observations. The Manhattan Project, of necessity, could not operate like that. Its activities were totally secret. The results of the speeded-up experimental work could not be published in the scholarly literature; the work could not be discussed at international conferences; other laboratories couldn't attempt to replicate the findings of researchers unless they were inside the tent. This secrecy rankled many scientists, and some refused to accept it. The more extreme became atomic spies. Others, such as Ernest Titterton, went the other way. Titterton relished secret work, as his later behaviour in Australia abundantly demonstrated.

Professor William Penney, another Manhattan Project physicist, was likewise comfortable with secrecy. His extensive background in secret wartime explosives and atomic weapons research equipped him to head the British nuclear tests in Australia. Penney was part of the small team who selected the targets for the Manhattan bombs, surely an onerous responsibility for a donnish mathematical physicist. He also visited Hiroshima after the bomb was dropped and conducted numerous scientific measurements on the ground.

In short, the Manhattan Project trained the men who later made the British bomb and brought it to the Australian desert. The huge covert project also taught them to keep their knowledge close. This ability to keep atomic secrets meant going against their scientific training. However, there is no reason to believe that these men (they were overwhelmingly men) were not sincere in their belief that the future safety of the world depended upon their ability to quietly and methodically change the nature of warfare.

The Trinity test of 16 July 1945 at Alamogordo in the New Mexico desert was a bittersweet moment for the science of physics. On the one hand, the brilliance of the thinkers engaged by the knotty problems presented by the bomb project prevailed. These were among the best minds of their generation, and, collectively, they moved nuclear physics and technology to a new realm. On the other hand, they unleashed a monster, and no-one was better placed than they were to understand this brute fact. While harnessing fundamental physical forces had undoubtedly given the whole enterprise a feeling of great adventure, the sobering reality hit when they saw the tangible evidence of their success in the form of a billowing mushroom-shaped cloud.

Oppenheimer, who had a literary bent, is said to have drawn inspiration from a John Donne poem to name the test Trinity. The plutonium bomb, nicknamed the gadget, and fundamentally the same as the weapon dropped a few weeks later on Nagasaki, produced the explosive force of 20 kilotonnes of conventional TNT. A select group of observers, including Oppenheimer and General Groves, were positioned about 32 kilometres from the device for the 6 am test. While some feared the device would fizzle (bets were taken on the outcome and fizzle was an option), instead it rose dazzlingly from the desert plain to create an awe-inspiring mushroom cloud that climbed upwards over 12 kilometres. A thump on the earth was felt by an oblivious civilian population in a 160-kilometre radius of the test site. Oppenheimer spoke his famous lines quoting Vishnu, destroyer of worlds, and recalled later in a haunting recorded interview, 'We knew the world would not be the same. A few people laughed, a few people cried. Most people were silent'. The atmosphere of the protective bunker is almost palpable in these words.

A few weeks later, on 6 August, people died in their tens of thousands because of this great leap forwards in nuclear physics. President Harry Truman said in his statement to a stunned American population immediately after the world's first A-bomb was dropped, 'It is an atomic bomb. It is a harnessing of the basic power of the Universe'. He continued, 'We have spent two billion dollars on the greatest scientific gamble in history – and won'.

The bomb dropped on Hiroshima was a crude nuclear weapon, codenamed Little Boy, based upon uranium-235 and using a technique known as gun technology that quickly became obsolete. It involved driving a cylinder of uranium-235 into the centre of another cylinder of the same substance with a hole in it. The bomb dropped on Nagasaki was nicknamed Fat Man and operated quite differently using plutonium. The basic idea of Fat Man (and indeed of the gadget) was to jam two half-spheres of plutonium together by detonating high explosives in a small space, initiating a chain reaction in which neutrons split the atoms and released energy. Fat Man was superior to Little Boy in design but had some drawbacks. In particular the weight of the conventional explosive needed to initiate the reaction made the bomb much bigger than Little Boy – hence 'Fat Man' – which presented logistical difficulties. The aircraft that dropped the Hiroshima bomb, *Enola Gay*, could not drop Fat Man. Instead, a specially modified B-29 called *Bockscar* was used for the task.

Some Manhattan scientists, notably the American Ernest Lawrence, argued for an eye-opening but non-lethal 'demonstration' of the weapon to the Japanese, rather than using it on human targets. The idea was dismissed. Washington agreed with the Scientific Advisory Committee: the shock value would be lost if the weapon was not used for real. Arguments about the ethics of that decision continue today.

In the final months leading up to the Trinity test, Winston Churchill wanted to ensure that the collaboration would continue after the war ended. Now that the idea of a fission bomb was becoming reality, with crucial input from both sides of the Atlantic, such a collaboration seemed both likely and desirable. The Hyde Park Agreement struck between Churchill and Roosevelt on 19 September 1944 said in part, 'Full collaboration between the United States and the British Government in developing tube alloys for military and commercial purposes should continue after the defeat of Japan unless and until terminated by joint agreement'. In fact, soon after the defeat of Japan, the Americans changed their minds.

After Hiroshima and Nagasaki, the world collectively took a deep breath. Atomic bomb research stopped abruptly, and bomb-making expertise dispersed for a short time as the Manhattan Project scientists and technologists went back to where they came from. While Little Boy and Fat Man had worked as they were designed to do, they were not well-developed bombs, and nuclear weapons needed more research, testing and refinement. But for a moment, everyone involved stopped to take stock, while the shock waves from the war in general, and the atomic weapons in particular, ebbed away. What had been wrought was so world-changing that for a while those involved did not know what to do. The genie was out of the bottle. During this time, both the UK and the US held talks with the United Nations (established in October 1945) to try to formulate a way that nuclear weaponry and energy could be harnessed without sparking an unstoppable nuclear arms race. The talks were unsuccessful and an arms race soon began.

The US gathered its thoughts on the existential issue of nuclear warfare and pondered the consequences of rapid atomic weapons development. One of its first postwar actions was to drive its atomic weapons allies away. The McMahon Act (known officially in the US as the Atomic Energy Act) was the result of this period of postwar reflection. This new American law, which banned collaboration on nuclear weapons development, took Britain by surprise and created a range of problems that the nation had not seen coming. Indeed, Britain saw the McMahon Act as a betrayal by the Americans, after Britain had handed over so much expertise during the war. Suspicions arose that the McMahon Act was a *commercial* decision, attempting to corner a lucrative new market in weaponry and energy. British know-how, combined with that of British-based European refugees who had escaped from the Nazis, made the atomic bomb possible. Suddenly Britain was elbowed out of the nuclear game. The country was displeased and wrong-footed.

The catalyst for this sudden and brutal excision of British science from the US atomic weapons program went by the name of the atomic spy Alan Nunn May. His name is inextricably linked with the events that later unfolded in Australia because in a sense he caused them to happen. British physics contributed most of the important atomic spies, although the US had its own too, including the prodigy Ted Hall, and the husband and wife Julius and Ethel Rosenberg, who were both executed for espionage. Nunn May was the first atomic spy to be revealed. His exposure further hurt an already fractious UK–US partnership on weapons development and set it back many years.

Alan Nunn May was a physicist, one of the young intellectuals at Cambridge in the 1930s tempted then seduced by the communists. He was briefly a contemporary of Donald Maclean, one of the renowned Cambridge spies and part of the Philby, Burgess, Maclean and Blunt circle, possibly Britain's most famous and romanticised spies. Nunn May graduated from Cambridge with an honours degree in physics then went to King's College in London to study for his doctorate and teach. While lecturing at King's, he joined a Communist Party group.

The balding, moustachioed and rather nondescript-looking Nunn May was not a particularly rabid or passionate party member. In fact, he had allowed his party membership to lapse by the time World War II began, when he was working on a secret British project to develop radar (incidentally, one of the other great technological feats achieved in the heat of war). In 1942, soon after the possibilities of splitting the uranium atom became known, Nunn May joined a team of Cambridge scientists who, as part of the Manhattan Project, were examining the technicalities of building an atomic reactor.

At that stage, Canada was becoming an important partner in the secret project to build an atom bomb. Canada mined some of the uranium ore (the rest came mostly from the Congo) and also extracted the uranium-235 and manufactured the plutonium needed to build the bombs. A vast technological effort was required to deliver the materials required. A fateful decision was made to transfer Nunn May to Montreal, where a Soviet spy ring was becoming active. The British physicist, already sympathetic to the communist cause, was recruited there by Soviet military intelligence.

As the Manhattan Project approached its final goal of a workable bomb, aspects of the massive scientific and technological push were being wound back. In July 1945, just weeks before the bomb was dropped on Hiroshima, Nunn May told his Soviet handler that he was about to go home. Moscow wanted to maximise its return on an asset. What could Nunn May give them that would make a lasting contribution? On 9 July 1945, the British physicist passed small amounts of enriched uranium (that is, uranium with a high proportion of the isotope uranium-235) to his handler. He later provided technical details of the bomb dropped on Hiroshima. In return, he received US$200 and a bottle of whiskey. He was never in it for the money, as he made clear at the end of his long life (he died in 2003 aged 91). He was a physicist who believed that knowledge should be shared, and that the secrecy around atomic weapons development was anathema to science.

Shortly after the war ended, Igor Gouzenko, a lieutenant in the Soviet military intelligence agency and cipher clerk at the Soviet Embassy in Ottawa, defected with documents that contained extensive details about Soviet agents, including Dr Nunn May. By then, Nunn May had returned to Britain, where he was arrested and put on trial. The day before his 35th birthday, in May 1946, he was sentenced to 10 years in prison. He was released in 1952, after serving just over six years. When Nunn May died in 2003, the *New York Times* quoted him as saying in old age, 'The whole affair was extremely painful to me, and I only embarked on it because I felt this was a contribution I could make to the safety of mankind. I certainly did not do it for gain'. The idea of the greater good was a common thread through much atomic espionage.

The arrest of Alan Nunn May confirmed American suspicions that the British were not to be trusted in the secret development of nuclear weapons. This, no doubt, was only worsened by the revelations that Nunn May had been investigated and cleared by MI5, the British security service. The Americans were, and remain, skittish regarding security issues. They didn't need much encouragement to form the view that other nations were lax in their security. They felt the same way about the Australians.

At the end of World War II, Britain was severely depleted in every way: emotionally, physically, militarily, socially and economically. The country had used all its resources fighting a desperate campaign against Nazi Germany. Rationing stayed in place well into the 1950s. Britain was broke. Yet she chose to go down the atomic weaponry path, to create a hugely expensive nuclear arsenal. Despite the Hyde Park Agreement, the McMahon Act forced Britain to do this without the valuable assistance of the US. Inevitably, this decision to become a nuclear power diverted economic resources away from postwar reconstruction and from a civilian population desperate for relief from the austerity of the war years. This was a significant choice.

The decision to become nuclear armed was not welcomed by all in the UK. One of the major critics was the chief scientific adviser to the Ministry of Defence Sir Henry Tizard, who famously said in 1949, after planning for the British bomb had begun:

> We persist in regarding ourselves as a Great Power, capable of everything and only temporarily handicapped by economic difficulties. We are not a great power and never will be again. We are a great nation, but if we continue to behave like a Great Power we shall soon cease to be a great nation. Let us take warning from the fate of the Great Powers of the past and not burst ourselves with pride.

Tizard, a chemistry researcher by training, had been involved with the Tube Alloys Directorate and was well versed in what was required to develop nuclear weaponry. Indeed, Tizard was among those who set up the Maud Committee. His view was not ill informed; he well understood what nuclear weaponry would cost the nation.

Nevertheless, Britain made a massive commitment at a time of great austerity. The secret decisions to build a nuclear arsenal were accompanied by related decisions to pursue nuclear energy; the two were intertwined. The country set up three new installations for the task: the Atomic Energy Research Establishment (AERE) at Harwell, near Oxford, which housed laboratories for research into both military and civilian uses of nuclear energy; the Windscale facility (later renamed Sellafield) in Cumbria, to manufacture the required radioactive material, including plutonium; and a research and development organisation that came to be called the Atomic Weapons Research Establishment at Aldermaston, to build the bombs and oversee weapons testing.

The eminent physicist John (later Sir John) Cockcroft headed the research establishment at Harwell. Cockcroft was already well known in the world of science. In 1932 with Ernest Watson he had split an atom by artificial means for the first time, an achievement marked by a Nobel Prize in 1951. The pair split the nuclei of lithium atoms using a beam of protons, demonstrating that nuclear fission was possible. During 1940 and 1941 Cockcroft had been a member of the Maud Committee that had proved the feasibility of an atomic bomb, and at the end of 1943 he had taken over as director of the Anglo–Canadian–French atomic team at Montreal. The translation of scientific research into practical engineering projects was a feature of Cockcroft's period as director of the Canadian project. Ernest Titterton went to work with Cockroft at Harwell after the war, as part of a physics research group, where he carried out fundamental studies of nuclear fission reactions. Titterton also acted as a consultant to the AWRE and established important contacts there. During this time Titterton published 28 papers. Then he packed his bags and headed to Australia in the early 1950s to begin an even more prominent phase of his career.

The task of overseeing the creation of postwar plutonium fell to the brilliant engineer Christopher Hinton, who was responsible for the design, construction and operation of plants to produce fissile material. He created the plutonium at Windscale for the British bombs. William Penney was appointed to head up the atomic weaponry research and development effort. Before his work on the Manhattan Project, Penney was a veteran of the Tube Alloys Directorate. According to Margaret Gowing, 'In 1946, as Cockcroft, Penney and Hinton took up their posts, it seemed that the special qualities and past experience of all three had combined to produce a very rare situation: the right men all arriving in the right jobs at the right moment'.

As Britain saw the pivot of world power shift towards the US, where the doomsday weapon had been made real, its leaders had a decision to make. Would Britain cede technological supremacy to the US (and, all too soon, the USSR), or would it match the rising superpowers bomb for bomb? The answer was not long in coming, although the public did not know it for some time. The British Government made its first step on the path to its own 'nuclear deterrent' in August 1945, shortly after the full extent of the unleashed atom was felt around the world. On 29 August, a circle of six UK government ministers who met in secret and called themselves GEN.75 started to consider a nuclear future (this was standard nomenclature for Cabinet committees, which were prefaced either GEN or MISC). Among other things, GEN.75 led directly to the formation of the Advisory Committee on Atomic Energy. The committee was the successor of the wartime Tube Alloys Consultative Council, a group of ministers and scientists who had advised the government on issues relating to the work being done by the Tube Alloys Directorate. Tube Alloys lived on under a new name.

Realism and idealism competed at this time in British nuclear planning. As scholar Susanna Schrafstetter put it, 'Britain was assembling a military nuclear programme while, at the same time, politicians were drafting lofty schemes to avoid nuclear war'. They wanted to head off a nuclear arms race and even considered sharing nuclear secrets with the USSR. But the reality of postwar geopolitics, and a desire to halt the country's international decline into irrelevance, focused minds instead on the challenging problem of how to simultaneously join the race and decry it.

The official decision to create a British bomb was taken in January 1947. An earlier passionate discussion within the tight circle of GEN.75 on 25 October 1946 that had essentially, though informally, set it on this fateful path could have gone either way. Foreign Minister Ernest Bevin arrived late to find fellow committee members about to decide against a British bomb and tipped the discussion in the opposite direction. By the end of the meeting, fired up by Bevin's rhetoric, they were all agreed. Britain would build its own bomb. In a foretaste of decisions about testing the bomb in Australia, the six GEN.75 ministers did not take the matter to Cabinet for discussion. Clement Attlee, the postwar prime minister, set up the short-lived GEN.163 committee on 8 January 1947 with the specific, though secret, task of building a British A-bomb. At a planning meeting held in June 1947 in the library at the Woolwich Arsenal, the practical matters associated with building Blue Danube, the first operational British nuclear bomb, began.

A grim atomic era calculus led the UK to determine that containing the USSR would require 1000 British nuclear weapons, a huge undertaking. The more enthusiastic among the backers of the British bomb believed it was possible. But building the British bomb was not a fast process. The committee drew on the expertise developed through the Maud Committee and Tube Alloys, and on the knowledge brought back to the country from the Manhattan Project. Ernest Bevin was particularly adamant that the snub by the Americans would not leave Britain out in the cold

While Alan Nunn May's arrest and imprisonment had significant consequences, they did not end the nuclear spy scandal. In 1950, another – much more important – physicist, Klaus Fuchs, faced the same fate. Fuchs, also a communist, was a brilliant scientist who was central to the Manhattan Project and a close colleague of Penney. He was arrested and served nine years in jail (Penney visited him there). In his confession he said that he had committed espionage 'in the name of historical determinism'. Fuchs had been attracted to communist ideology in his home country, Germany, during the early days of the Nazi regime. He escaped to the UK in 1933 and studied at various British universities during the 1930s. For a brief time in 1940, he was interned as an enemy alien on the Isle of Man, and was moved to Canada before returning to Britain after the physicist Max Born petitioned for his release. He resumed his physics research, and his gifts as a physicist were quickly appreciated through his work, from May 1941, at the University of Birmingham. He supplied information to the Soviet Union from the time he joined Birmingham and worked under Rudolf Peierls, co-author of the Frisch–Peierls Memorandum. He went to the US in 1943 with the British mission, after the Maud Report had shown the way to a nuclear bomb. Around the same time he became a naturalised British citizen.

Fuchs worked initially in New York and later at Los Alamos as a theoretical physicist. He handed over a large amount of information to the Soviets about the forthcoming Trinity test. This essentially meant the bombs dropped on Japan were no surprise to Moscow. At the end of the war, Fuchs returned to Britain and was centrally involved in both nuclear energy and nuclear weapons development. He operated separately to Alan Nunn May and was not part of the same espionage ring, although they did share the same Soviet contact at one time.

When Nunn May was arrested, Fuchs was concerned enough to avoid using the London-based contact who was involved with Nunn May. He continued his espionage, though, passing secrets to a new contact during meetings at the Nagshead pub in London. Only then did he take payment. He accepted about £100 to help defray the costs involved in the extra security precautions he required in the light of the Nunn May exposure. He also believed that accepting some money would assure his new contact of this loyalty.

The first Soviet atomic weapon tested, in 1949, was similar in concept to the Fat Man weapon dropped on Nagasaki. It was made possible in large part by the hundreds of documents Fuchs had passed to his Soviet handlers. In recent years, more evidence has come to light that Fuchs also passed information about the much more powerful fusion weapons that followed. Many of those documents originated from William Penney, although there is no suggestion that Penney knew about Fuchs' espionage activities. Fuchs was probably the most damaging of all the nuclear spies. Unlike Nunn

May, Fuchs was part of an elite group within the Manhattan Project centrally involved in the bomb project. He also spied for longer and passed crucial information about fusion weapons.

The American spy Theodore (Ted) Hall was a scientific prodigy who graduated from Harvard at 18 and was only 19 when he joined the Manhattan Project. He also supplied technical information to the Soviets about the plutonium bomb dropped on Nagasaki. According to Hall's 1999 obituary, 'Of all the scientists, diplomats and others who passed atomic secrets to Moscow – Fuchs, Maclean, Nunn May … and the rest – it is likely that only Fuchs was more valuable [than Hall] to the Soviet bomb programme'. Like Nunn May and Fuchs, Hall did not like scientific secrecy. He believed that the world would be a better place if nuclear knowledge was shared. His contribution was more significant than that of the Cambridge spies.

The Cambridge set are the best known spies, but in a sense the least damaging. They were not scientists and so were unable to share physics like Nunn May, Fuchs and Hall. Kim Philby, Donald Maclean, Guy Burgess and Anthony Blunt passed a variety of non-technical nuclear secrets to the Soviet Union, as well as polit-ical and tactical information that kept the USSR informed about the strength of the US atomic arsenal. In his 1980s book *Spycatcher*, Peter Wright, a former agent of the British spy service MI5, alleged that Roger Hollis, who headed MI5 between 1956 and 1965, was a 'fifth man' Soviet agent. (Hollis visited Australia in the late 1940s to investigate the allegations of espionage that became the impetus for creating ASIO.) British journalist Chapman Pincher had earlier made the same allegations. The first Cambridge spy was uncovered just after Fuchs, in 1951, although it took many years for them all to be outed. To this day, doubt remains about whether or not Hollis was a spy too.

As the spies were uncovered, the US retreated behind the McMahon Act. But in the UK, work on the bomb proceeded. The GEN.163 committee set up a new organisation called Basic High

Explosive Research and appointed the Manhattan veteran Penney to head it. Penney had recently had a stint at Bikini Atoll, working with the Americans on their postwar bomb (as had Klaus Fuchs), just before the McMahon Act became law. His name is synonymous with the British nuclear tests in Australia, and his role will be examined in more detail later. He was the best and most logical person to take the British nuclear weapons development campaign forwards. GEN.163 clearly set his pathway to Maralinga.

Penney continued to work with his old Manhattan confreres, including Titterton and (fatefully) Fuchs. The project name was soon changed to High Explosive Research (HER), continuing the British tradition of choosing banal names to deter interest. The quest to build a British A-bomb was first announced to the public on 12 May 1948. HER researchers initially worked at various laboratories and test facilities in Kent, Essex and Suffolk, until they moved to a former airfield in Berkshire called Aldermaston. The atomic weapons development and testing activities in Australia were managed from Aldermaston. Some work was also carried out at Fort Halstead and at the Woolwich Arsenal, where Penney was based in those early days. HER was renamed the Atomic Weapons Research Establishment (AWRE) in 1950.

One of Penney's many lasting contributions was Blue Danube, Britain's first tactical nuclear weapon. The design of the weapon was solid, sound and (according to many, including official historian Margaret Gowing) better than comparable American bombs at the time. A fission weapon powered by plutonium, it was similar in many ways to the Fat Man device detonated over Nagasaki. However, its unique design features made it more efficient and easier to assemble, control, aim and store.

Penney brought an excellent problem-solving mind to the task and oversaw four secret research and development groups that worked on key aspects of the bomb's design. The team also worked with a wide range of commercial engineering contractors and manufacturers throughout England and Wales. They managed to maintain a secret operation despite the many people involved. As Jonathan Aylen put it:

> Blue Danube's practical development was a product of a wider 'warfare state' of Cold War Britain, not just the product of a few boffins. In truth, the first atomic bomb was designed in suburban centres and built in ordinary factories down prosaic back streets by regular workers – men and women – in towns across industrial Britain.

There is something ineffably British in this process, just as when the Maud Committee used laboratory word games to come up with the most impenetrable codenames for work done in plain sight.

So Blue Danube was created in the dingy factories and hastily built research and development laboratories of postwar Britain. It came into being in a roundabout way, untested, during extraordinary and tumultuous times. The leaders of the project to create it, including the brilliant mathematical physicist Penney, were confident that it would work. But confidence is not enough. Weapons must always be tested. As the HER project proceeded, and the scientists conveyed their reports to their political masters, the next step was clear. Blue Danube would have to be exploded. Since the nuclear spies and the McMahon Act had put Nevada and New Mexico off limits, the question arose: where could the fledgling British nuclear deterrent be tested?

3

Monte Bello
and Emu Field

We've got to have this thing over here whatever it costs ...
We've got to have the bloody Union Jack on top of it.

Ernest Bevin, UK foreign secretary, at GEN.75, 1946.

I saw this ominous black atom bomb sitting on the top of the tower,
bolted in place, winched up to the top on guiderails that I'd laid
in. And I saw a crow sitting on it. I said to the crow, 'if I were
you, I'd shift'.

Len Beadell, legendary surveyor of Central Australia, speaking in
1991 about the first Emu Field atomic test held in 1953.

Australia in the 1950s was not a country casting a critical eye
at the development of, or consequences arising from, nuclear testing.
There is little evidence of the doubts and fears which American
scientists who pioneered the nuclear project expressed.

Paul Malone and Howard Conkey, *Canberra Times*, 1984.

The McMahon Act was a terrible blow to William Penney and his team at Aldermaston. They had designed a technologically advanced atomic weapon, but they couldn't test it at the American sites. As Penney said later in his statement to the Royal Commission, 'I consistently took the view that I would prefer to use the existing American facilities, either in Nevada or the Pacific'. He was most at home in Nevada, where he had earned his nuclear weapons stripes in the inner circle of the Manhattan Project, and where he could work among trusted colleagues.

The decision to bar outsiders looked permanent in 1946, and the British had no time to lose. Some behind-the-scenes negotiation for ongoing low-key co-operation had taken place, albeit with some onerous conditions applied by the Americans, but all hope collapsed in 1950 when Klaus Fuchs was exposed as an atomic spy. The US could not afford to dice any further with disaffected scientists looking to make a high-minded ideological point. The decision was final. The Americans amended the Act in 1958, well after the first British atomic device had been tested, and invited them back. But until then Britain was on her own.

Penney assumed the next option would be Canada, which had been directly involved in the Manhattan Project. He was familiar with its expertise and infrastructure and conducted a feasibility study there in the late 1940s. In addition to a test site, he was looking for a long-term collaboration. The aim was to conduct the first British test in the northern summer of 1952. The location needed to be quite particular, because Penney was keen to test the effects of detonating a weapon based on his Blue Danube design aboard a ship. He wanted to analyse the minutiae of a phenomenon known as base surge, which involved a large upward movement of radioactive water into the air, when the products of a nuclear explosion mixed with fine droplets of water. Disappointingly, none of the seven ports he examined was suitable for a shipboard detonation. Nevertheless, Canada remained on the drawing board.

According to official British historian Margaret Gowing, Penney envisaged one or two annual trials for several years, at a site generously staffed and equipped with instruments. He recommended 200 scientists, 50 technicians and 100 industrial workers, at huge potential cost, for the first trial. 'Most of the scientists would be provided by Britain, with help from the Canadians in chemical analysis and radiological safety, and most of the industrial workers by Canada; Canada would undertake the construction work; costs would be shared on an agreed basis.' Canada's wartime experience in contributing to the Manhattan Project provided a strong platform for a Canadian test program.

Penney had a variety of criteria he applied to the seven potential Canadian sites, mostly to do with climatic conditions, infrastructure and isolation. Churchill, Manitoba, on the west coast of Hudson Bay, seemed ideal, even though the shallowness of its port made base surge observations difficult or unlikely. Churchill is about 1600 kilometres north of the provincial capital of Winnipeg and extremely remote, not to mention unspeakably cold during a substantial part of the year. Today it is best known as a tourist destination to view polar bears in the wild. In his report, Penney acknowledged that the area would be contaminated by the proposed tests but described the land near Churchill as 'valueless'.

History shows that the British never did test a nuclear weapon in Canada. In fact, as scholars John Clearwater and David O'Brien noted, there is no evidence 'to show that the elaborate proposal, which included detailed plans for roads, barracks and other infrastructure, ever went to the ministerial level'. When the extent of the proposed contamination of the Churchill site became clear, the Canadians quietly shelved the idea. Again, Britain was on her own.

Meanwhile, Australia came up on the radar. This peaceful backwater of a country seemed a natural choice in many ways: developed, Western, a member of the Commonwealth and a former colony. It possessed huge potential sources of uranium, uninhabited islands and swathes of desert. At that moment in history, a noted Anglophile, Robert Menzies, headed its government.

The fact that Australia had virtually no background in matters nuclear, other than the largely expatriate contribution to physics research of Mark Oliphant and a few other scientists, was irrelevant. Britain would control the science. What they needed was space, not scientists and technologists. Prime Minister Attlee made the first approach, and Menzies was almost certainly taken by surprise. Nevertheless, when he received Attlee's top-secret cable on 16 September 1950, followed up by a telephone call, his instinctive eagerness deterred him from consulting colleagues. Australia's Monte Bello Islands had cropped up because Penney's interest in observing base surge had led to canvassing island locations. It rose to the top of the list when relations with the US over atomic matters deteriorated with the arrest of Fuchs, and the Canadians went cold on the sub-Arctic site at Churchill. Other Pacific island nations were also considered but discounted. Groote Eylandt in the Gulf of Carpentaria was briefly on the list but not pursued because of its meteorological conditions.

A swift survey to confirm Monte Bello's suitability for the proposed historic first British test followed. The top-secret Operation Epicure was overseen by Major General AJH Cassels in the UK Services Liaison Staff Office in Melbourne and by Sir Frederick Shedden, the secretary of the Australian Department of Defence. Their cover story was that they were investigating the feasibility of extending the UK rocket testing project. An ASIO officer joined the survey, the first time the new domestic spy service was involved in the push for atomic weapons. The Royal Australian Navy provided HMAS *Karangi* as the survey ship and sent HMAS *Warrego* to prepare a detailed chart of the complicated maritime features of the island group, since little information was available at the time. The project used the term Western Islands, rather than Monte Bello, to maintain secrecy. As well as charting the islands, the crew of the *Warrego* found an immensely rich natural environment, which they outlined in their report. Many of their biological samples eventually found their way to the British Museum in London.

Epicure was swift indeed. *Karangi* had done her work by 27 November 1950 and steamed back to Fremantle, bearing a trove of information. Data included depth soundings and information on the tides and the winds. The Epicure results were submitted for review to the AWRE and government representatives in London, where they found favour. In February 1951 the British chiefs of staff agreed to a shipborne atomic test at the Monte Bello Islands. Epicure had established that a maritime explosion and base surge could be tested there.

In fact, Epicure was so quick that some people, particularly the slow and steady Shedden, were concerned about making serious decisions with long-term consequences under undue time pressure. The survey confirmed that October 1952 would be the best time for the test – indeed, October was probably the only feasible time of the year. Attlee confirmed the details with Menzies on 27 March 1951. Attlee added, coyly, 'We can settle later the details of finance and machinery'.

As it happens, these details were never properly settled. The Australians behaved as though agreeing to pay for expensive parts of the test operations in Australia would be something of a bargaining chip to obtain knowledge from the British tests, but did little to actually deploy this supposed advantage. Gowing later wrote, with magnificent understatement, 'The Australians agreed to this without striking a hard bargain over technical collaboration'. And so an initially secret deal was agreed, well before most members of the Menzies government, let alone the Australian people, knew. The massive operation had less than 18 months to be organised, and the Monte Bello Islands were a world away from the British Isles.

Penney, who ventured forth from his base at Aldermaston to take charge, was in his early 40s when he directed Operation Hurricane at the remote Monte Bello Islands. Born on 24 June 1909 in Gibraltar, he was educated in England and the US and held PhDs from both Imperial College London and Cambridge. His expertise in blast waves from explosions led him to the Manhattan Project, which in turn, he told the Royal Commission, guided his work on the nuclear tests in Australia. He led the Australian atomic test series for all the major trials, although from 1954 he became increasingly involved with the testing of thermonuclear weapons in the Pacific and delegated some Australian tasks to AWRE deputy head William (later Sir William) Cook.

Penney remained director of the AWRE until 1959 and later became chair of the UK Atomic Energy Authority. He was a fine-looking man, tall with broad shoulders and a strong, square face that sported thick-rimmed glasses. He had a smooth, posh accent that often seemed to be on the brink of sophisticated merriment. He was also dignified and likeable. In fact, even royal commissioner Jim McClelland could not despise him in the way he despised Ernest Titterton, though he hated everything that Penney presided over in Australia. In an obituary in 1991, Penney was described as 'a friendly, undevious and usually humorous man' and 'a shrewd administrator and good judge of people'. For his work on developing the British bomb and beyond he was knighted, and later made a life peer, taking the title Baron Penney. To many who knew him, though, he was simply Bill – or indeed Buffalo Bill, after Maralinga's Buffalo series. When he fronted the Royal Commission in London in February 1985, he had changed physically, with one newspaper describing him as 'a small round man with round glasses and wispy white hair, wearing crumpled tweed clothes under an equally crumpled coat'.

Another towering figure in this saga, who made the British nuclear tests possible with a stroke of his pen, was the Australian prime minister. Robert Menzies, an unashamed Anglophile, began his second tenure as prime minister in 1949, and it lasted 16 years. In fact, his enthusiastic co-operation with Britain's nuclear testing program was a defining characteristic of his prime ministership. It fitted in with his overall desire to reinvigorate the relationship between Australia and the UK, which had been bruised in some ways by World War II and the Depression.

His enthusiasm for nuclear testing was not considered strange at the time. Despite the initial shock at the destruction of Hiroshima and Nagasaki, many people in the West saw atomic weaponry and energy as positive and forward-looking developments, a view encouraged by US and British propaganda. Also, if Britain was doing it then that was fine by Australia. As secretary of the AWTSC John Moroney remarked in 1993, the times were different. The 'closeness and strength of feeling between' Britain and Australia, said Moroney, 'was a very tangible thing then, but virtually incomprehensible to many now'. To many Australians, Britain was the 'Mother Country', and Australians visiting the UK, even for the first time, talked about 'returning home'. While some segments of Australian society were resentful that Churchill had tried to stop Australian troops from leaving the Middle East to defend Singapore and Papua New Guinea from the Japanese during the war, the majority of Australians saw loyalty to Great Britain as a natural part of the order of things.

This close relationship helped to secure the original agreement and set in train a long series of events still not completely resolved. The connection between the UK and Australia altered during the nuclear tests saga and its aftermath. The initial Australian willingness to agree to British requests for weapons testing became less ardent over the 11 years they lasted. Even at the beginning, though, Menzies knew that atomic testing could be politically difficult. He asked Attlee to delay finalising the official arrangements for the first British nuclear test until after the Australian election of May 1951 that returned the Menzies government for its second term. Such was not the fate of Clement Attlee, who lost his election in October.

Momentum was not lost, however. British wartime leader Winston Churchill, who was returned to power, was well onside with the idea of a British nuclear deterrent. When he learned, to his surprise, just how far advanced plans were, he was impressed and enthusiastic. A strong influence was his chief adviser Lord Cherwell, who barracked relentlessly for the UK to build its own A-bomb. To do this, Britain finally had to face facts and give up on the US. Plans for

Hurricane were formally agreed between the UK and Australia on 27 December 1951.

Menzies kept much information about the British plans to himself and the small group of advisers and public servants such as Frederick Shedden who were closely involved. In the earliest stages, only Menzies, the minister for Defence Philip McBride and the treasurer Arthur Fadden knew. After striking the agreement to test the British bomb in Australia, Menzies constructed a formidable apparatus of secrecy. Most of his Cabinet still knew little or nothing, even after the Churchill ascendancy. The meetings were top-secret, and only limited information was shared outside the tight circle of insiders.

Meanwhile, much furtive activity was underway. The Monte Bello Islands had no infrastructure, so a massive operation began to deliver all the paraphernalia of a nuclear test and its associated scientific studies. The movement of navy vessels from Fremantle created some speculation though. As well as *Karangi* and *Warrego*, the navy deployed HMAS *Mildura* to assist with setting up the site. *Karangi* ferried heavy equipment, including a prefabricated hut, two 25-tonne bulldozers, a grader, a number of tip trucks, several electrical generating sets, twenty 1.8-kilolitre water tanks, a mobile transmitter and receiver, and the plant needed to establish refrigerators.

The whole thing took on the appearance of a major naval operation, which proceeded without any public acknowledgment. The grapevine along the Western Australian coast became progressively more active as sharp-eyed locals put two and two together, and the odd newspaper story also hinted at what was afoot. However, and remarkably, secrecy was maintained to the satisfaction of the authorities. When the Australian Government finally announced to the public that Operation Hurricane was about to begin, the media release congratulated all concerned: 'Indeed, the degree to which secrecy about the really vital matters has been preserved is a splendid tribute not only to the security officials but also to the loyalty, integrity and sense of respect displayed by everyone concerned in the vast project'.

Soon after agreeing to provide Monte Bello to the British for atomic testing, the Australian Government realised they needed to formally exclude people from the test site. Although Monte Bello was remote and rarely visited, naval activity in the area could provoke curiosity. The odd nosy boatie might just turn his rudder in the general direction to have a bit of a look. To head off this possibility, the *Defence (Special Undertakings) Act 1952* passed through federal parliament, making unauthorised entry illegal. At first, the Act embraced only Monte Bello, but it later covered Emu Field and Maralinga (and later still the US signals monitoring station at Pine Gap), effectively excluding citizens by law from portions of Australian territory. The prohibited area extended over all the Monte Bello Islands and their surrounding waters, and south to Barrow Island. The Act was passed quickly in June 1952 – so quickly, in fact, that it was alleged that Australia had accidentally prohibited areas outside the country's territorial limit, over which it had no jurisdiction. Both the UK Foreign Office and the Commonwealth Relations Office (CRO) believed that it extended at least 32 kilometres outside Australia's territory. The UK had been burned with disputes in the past over maritime boundaries and wanted the legislation amended to avoid future embarrassment. Despite their concern, it wasn't.

So the inexorable process of turning Australia into a central player in the accelerating international nuclear weapons proliferation began with a combination of speed and a small but telling element of ineptness. The test location was an unassuming group of islands in a remote part of the nation, unknown to most Australians. Monte Bello is an archipelago of 174 small islands about 130 kilometres from the strikingly beautiful Pilbara coast, where the mining town of Karratha was established in the 1960s. The total land area of the islands is only 22 square kilometres, and they lie 20 kilometres north of Western Australia's second largest island, Barrow Island, best known for its high conservation value and fossil fuel reserves. The limestone and sand Monte Bello Islands are mostly low in the water and covered with scrubby vegetation. The British chose Trimouille Island for the first test in 1952, then Alpha and Trimouille for the two Mosaic tests in 1956. Three atomic devices were tested at Monte Bello, one of which was so big that it sent a radioactive cloud over the entire Australian continent and created a cloud of suspicion about its true nature that took many years to clear.

Penney was happy enough with Monte Bello, under the circumstances. It was no better than third choice (after America and Canada), but it would do, and time was getting away. He selected the UK rocket scientist Dr LC Tyte as planning director for Hurricane, with the ballistics expert Charles Adams as his deputy. It was essentially a maritime operation, and Rear Admiral Arthur Torlesse from the Royal Navy was appointed to command the ships, while the Royal Australian Navy supplied patrol ships. Australia contributed extensively to the preparations. When the British scientific staff arrived at Monte Bello, they were amazed by the systems of roads on both Trimouille Island and nearby Hermite Island, where the field laboratories were built, as well as the network of buildings and laboratories.

Operation Hurricane was specifically designed to detonate a nuclear device from 2.7 metres below the waterline in the hold of a ship. Britain was a maritime nation with many ports. The British authorities were keen to find out what would happen if an atomic bomb was detonated in a port, a scenario that would radioactively contaminate a large volume of water. Not even the Americans had done this sort of test before. Highly contaminated water would rise up in a column upon detonation, and it would have to come down again. The British readied themselves to measure and analyse when, where and how this would happen. HMS *Plym*, a relatively new 1390-tonne frigate that was surplus to requirements after the war, steamed all the way from the UK carrying in its belly the framework for an atomic device. *Plym* was part of a Royal Navy flotilla that included four other vessels. She came the long way, around the Cape of Good Hope, because the narrow Suez Canal was deemed too risky. The plutonium heart of the device came separately, by seaplane, via the Mediterranean, the Middle East and Asia. At the end of her long voyage, the doomed *Plym* was anchored in 12 metres of water about a kilometre off the west coast of Trimouille Island.

The atomic device was detonated at 8 am on 3 October on Penney's orders. Hurricane was a plutonium-based implosion weapon, using 7 kilograms of plutonium manufactured at Windscale in the UK. Penney had started to design it in 1947, when he joined the GEN.163 committee. The device was expected to have an explosive yield of 30 kilotonnes, although in the event its yield was 25 kilotonnes. Penney and Torlesse watched anxiously from the Operation Hurricane command ship aircraft carrier HMS *Campania*, anchored about 5 nautical miles to the southwest. The other ships were *Narvik*, *Zeebrugge* and *Tracker*. The Australian party was watching from *Narvik*.

What Penney saw through his binoculars must have filled him with a mixture of relief and awe. His atomic device worked. Observations showed that an atomic fireball emerged 23 microseconds after detonation, almost obliterating the frigate. A vast column of water, approximately 1100 metres across and 170 metres high, rose up alongside the atomic cloud. Twenty-four hours after the explosion radioactivity was detected at an altitude of 3000 metres between Port Hedland and Broome. The mushroom cloud was a kind of deformed 'S' shape, notably different from the normal symmetrical shape.

Aside from the atomic weapon test itself, a huge variety of experiments was conducted to examine the effects of an atomic blast on equipment and infrastructure. These were known under the collective term target response studies and were of both military and civilian interest. For example, the authorities sought to test how well protected humans might be if they sheltered in various kinds of trenches and bunkers. Tests were also conducted on the personnel who entered the contaminated areas after the blast. As the director's report on Operation Hurricane stated, 'Some men needed 5 showers to pass the (very low) tolerance at final monitoring'. Guinea pigs and rabbits were placed in the blast zone for zoological testing. These tests clearly showed that the biggest danger came from 'invisible' fallout rather than 'black' fallout, pointing to an insidious risk.

The Hurricane test showed that Penney's design worked when delivered in the hold of a ship in port, but this was a long way from being dropped from an aircraft, the most likely delivery system at the time. Many more tests were needed. After further tests during Operation Buffalo at Maralinga in 1956, the weapon acquired the name Blue Danube, and it subsequently became Britain's first operationally deployed nuclear weapon. Hurricane was an important step. It enabled Penney and colleagues to refine the design and also provided tangible evidence of the UK's (somewhat belated) return to strength and prestige after the depredations of the war. And, of course, most significantly Operation Hurricane raised the UK into the nuclear-armed club, where it remains to this day.

Howard Beale, the Australian minister for Supply, who soon became the public face of the tests in Australia, found out about the first test only a short time before it was scheduled. In the lead-up to the test program, Menzies acted unilaterally, and Beale was kept in the dark. Beale twice, unknowingly, misled parliament by saying that there were no plans for British tests in Australia, in June and October 1951. As Menzies' biographer AW Martin put it, 'Menzies' unquestioning acceptance of the British insistence on secrecy, while fitting with his current Cold War fears and appreciation of American attitudes, created some strange situations. One of the strangest was his refusal for many months to admit his Minister of Supply, Howard Beale, into the secret'.

Beale later admitted how he 'boiled and fumed at what I regarded as an insult', but he quickly became a strong public voice for the test program. Beale and his department provided the material for much of the media coverage of the tests until 1957. The scientists' task, including that of Australian scientists, he wrote, was 'to make sure that the tests were safely conducted, and it was my department's task to give all required assistance and to keep the public informed. When it was announced that the test would take place, there was little public anxiety; indeed there was some pride that Australia was to participate in this historic event'.

Media coverage of Operation Hurricane in October 1952 was the first opportunity for the test authorities to interact with the Australian media, and they showed inherent caution in all their dealings. At a conference in 2006, counsel assisting the Royal Commission into the British Nuclear Tests, Peter McClellan (not to be confused with commission chair, James McClelland), told a story about the media releases connected with Operation Hurricane. He claimed that before Hurricane 'three press releases were prepared. If the test was successful the announcement was straightforward – a glorious success. However, if it failed or partially failed an excuse had to be found. That excuse and the cables publishing it had been drafted long before Lord Penney gave the command to explode the bomb – and it would not have mattered if it reflected the real truth'.

Robust requests from media organisations for journalists to join the official party that witnessed the test were, after initial consideration, denied. Some media chose to circumvent the restrictions and set up their cameras at Mt Potter, 88 kilometres from the test site, from where they captured images of the explosion. These ran prominently in a number of newspapers – a clear sign that keeping the huge mushroom clouds out of the media was going to be impossible.

An editorial in the *West Australian* of 4 October 1952 was typical of the newspapers' response at the time: 'The real significance of the Monte Bello explosion lies at this moment ... in the simple fact that it occurred. It gives the world the indisputable proof that Britain has the material, the skill and the installations for the independent production of atomic weapons and that she will yield the initiative to none'. It concluded that 'the Monte Bello explosion reverberates with a vastly increased assurance of British Commonwealth power and defensive security'.

Hurricane was both a scientific and a propaganda success. Penney, suave and self-effacing, became a national hero and was knighted immediately. He was also given the green light to continue his test program in Australia. After Hurricane, Penney and his men departed the maritime environment for a place with even more challenges and difficulties – a remote South Australian desert site labelled Emu Field. This dot on the map was in the far northwestern reaches of the Woomera Prohibited Area that took up one-eighth of South Australia. The codename was Operation Totem. The main aim was to prepare a British atomic device for deployment aboard RAF V-bombers and to see if mass production using cheaper production methods was possible. The British had gathered a lot of data from Hurricane but did not yet have a compact and cost-effective bomb that could be dropped from a plane. Totem set out to fill the knowledge gaps.

The search for a mainland Australian site had begun before Hurricane. The Australian bushman and surveyor Len Beadell, who found the Emu Field site, was a noted raconteur and he loved to tell the story. He was way outback early in 1952 when he got an urgent call on his radio, a rare event. He was to return to Salisbury without delay; something big was afoot. Just getting to the Stuart Highway took him a week.

> When I finally got back to Adelaide they locked me up in a little tiny office. And six people were glaring at me; they drew the blinds and soldered the keyholes over. The chief security officer started off the conversation in the most friendly way I had heard in a long time, merely because I hadn't heard anyone for a long time. He said 'what we are going to tell you now is known to these six people and nobody else, and if it gets outside this room it will be one of us and we will find you and it will be a nine year jail sentence'. And I thought to myself 'I might even keep it to myself'. The chief scientist for the whole project [probably Alan Butement] carried on then and said 'what we are going

to tell you is that we are going to explode an atomic bomb in Australia and we want you to pick out a site.

Beadell was a natural choice. He had initially been employed by the Australian Government's Long Range Weapons Establishment (LRWE) at Salisbury near Adelaide in 1946 and had spent a number of years out bush, surveying the enormous Woomera rocket range. As part of that project, he had established the 'centreline' – a 3900-kilometre corridor across the Great Victoria, Gibson and Sandy deserts that did not contain any cities or towns. It did, of course, cut right across Aboriginal lands, but shamefully that was not a priority to governments at the time, and it did not seem to ruffle Beadell's laconic bushman persona either. Rockets were fired along the centreline for many hundreds of kilometres. Beadell said during a talk to a Rotary convention in 1991, 'That centreline, although I didn't know it at the time, was going to govern the whole of the future of Central Australia forever'.

It certainly governed the way both Emu Field and Maralinga were established. Both sites were to the southwest of the centreline, but Emu was much closer to it than Maralinga. Emu's extreme remoteness meant that all supplies were brought in by air, and its closeness to the centreline interfered with Woomera guided missile tests. This meant that Emu was never suitable in the long term for atomic weapons tests. However, Beadell fulfilled his brief to find a remote location within the Woomera Prohibited Area with terrain suitable for both an airstrip and a weapons test area. The site was remote all right; it ended up being too remote for the long haul. Still, for the first mainland tests in Australia the AWRE wanted to be well out bush.

The dramatic announcement inside the claustrophobic room when Beadell was told about the bomb started a five-month search across 48 000 square kilometres of harsh territory.

I finally found, purely by accident, a clay pan, on the other side of a sandhill. I went over the sandhill and there was this clay pan, a mile long, half a mile wide. One glance told me that you could land any weight aircraft we had at the time without any preparation – a perfect natural runway in the wilderness. I called this my base and I searched out in all directions in a radius of about 100 miles from this trying to find an area free of sandhills and mulga scrub, an atomic bomb site … would be some place where you could move about without having to climb up over sandhills and push through thick scrub. You'd need plenty of room for instrumentation to record the results from the bomb itself, a place where you would put a tent camp for people to live in, and you certainly wouldn't want solid mulga and sandhills. So one of the directions 40 miles away I did find somewhere where the sandhills diminished and disappeared, the mulga scrub thinned out to open saltbush paddocks and in-between were little clumps of mulga and I thought 'this is a perfect site for a bomb'.

The site chosen by Beadell was given the surveyors' designation of X200. It had to be approved by Penney, who saw it in September 1952, on his way to Operation Hurricane. Beadell was slightly awestruck by Penney, while still pulling his leg at every opportunity:

This man had a genius of a mathematical brain on a level with Albert Einstein. He was the director of the whole concern, director of the atomic weapons research establishment in the UK, and he was going to come to my camp, stop with me for two weeks while I entertained him and showed him my bomb site for Australia.

Beadell had organised the delivery of eight Land Rovers to the site under the most trying circumstances. When the two large aircraft carrying Penney, his associates and the chief scientist for the

Australian Department of Supply Alan Butement were due to land on the claypan, Beadell lined up the cars with their headlights on to guide them in.

Only the smallest number of people knew about plans for a mainland site. The pilots of the two aircraft had reluctantly been let in on the secret, but at this stage Beadell was one of the few people in the world who knew what X200 would become. Penney was a recognisable figure, so his trip to the centre of Australia was a highly secretive affair. After a 'beautiful landing' and an ice-breaking joke shared between Beadell and the British scientist over the state of Beadell's socks, the party had to make the arduous trip from the makeshift airstrip to the proposed testing site about 80 kilometres away. Beadell observed that it was not easy since none of the party had 'driven anywhere rougher than Piccadilly Circus'. The bushman had to edge them along painstakingly, over sandhill and through thick scrub.

> Before we left the edge of the clay pan we stopped to let them see the little pebbles around the edge of the clay pan, the formations, and one of them said 'what made the footprint in the wet sand, in the once-wet clay?' And I said 'oh, that's an emu's foot. When a bit of a shower came here once, an emu put his foot on the soft mud and it left an imprint. That's an emu's foot'. So we started to call this clay pan, 'the clay pan with the emu's foot mark on it'. Gradually that just became 'the emu clay pan', and the following year when the bombs did go off, it went around the world on the front page of every newspaper, 'Emu Field atomic test successful', and that's where the name Emu came from.

A safety assessment of the site was sent to the Australian scientist and defence scientific adviser Professor Leslie Martin (soon to be appointed head of the AWTSC) and also to Ernest Titterton in May 1953. Emu was a diabolical site. Apart from being remote, it was a hardship post where everything was difficult. There was little water and no infrastructure. Everything had to be landed on the claypan and driven through the scrub, including a Centurion tank. Beadell had to lay out the instrumentation and set it all up. It took him a year to lay out thousands of instruments arrayed around the Emu bomb site to record data from the tests.

Again the arrangements proceeded rather too speedily given their complexity and logistical difficulty. Totem was a comparative trial, and its two devices contained differing proportions in plutonium-240. Both devices were detonated from 30-metre steel towers. They were much smaller in yield than Hurricane's 25 kilotonnes: Totem 1 was 9 kilotonnes and Totem 2 was 7 kilotonnes.

Penney kept Totem under tight control, with little Australian input. The only exception was Ernest Titterton, now working in Australia, but always seen as essentially on the British side, with his Manhattan Project and Harwell credentials. Titterton was given access to documents that set out the firing conditions and predicted contamination for the Totem series, and generally drawn into aspects of planning for the series. He had been party to insider information on Hurricane, too, to a much greater extent than his Australian colleagues. Titterton was given the chance to revive some of his postwar Harwell research at Emu Field when he conducted some field experiments during the Totem series.

Beadell had a front-row seat for the first Emu Field test, Totem 1, on 15 October 1953.

> I was standing alongside Sir William Penney on a little rocky
> outcrop which I had shown him before when I first took him
> down. I was only joking, but I said to him 'it would be a nice
> place to watch the bomb go off' ... but that's where we were.
> The countdown got to minus 10, nine, and I said to him 'we're
> only 4 miles away from this' and he said 'oh, it will probably
> be all right'. And I said 'I'd planned on being 400 miles away'.
> He said, 'we'll go together anyway'. We had our backs to it,

and when it got to zero the whole of the world that we could see lit up in the most blinding orange flash that you couldn't describe on such a scale. It lit up the whole sky, obliterated the sun completely and disappeared over an 80 mile skyline in the distance – the whole of the sky. I could feel the heat of it on the back of my neck. We turned around to see what we had done. Sir William and I got into an aeroplane and flew over it to see what it looked like from the air. If I had had the advantage of reading about what atomic radiation does to people after a long period, I might not have been so keen on going on that flight … When we flew over it, all we could see was a half a mile diameter sheet of melted sand, and nothing else.

The aftermath was different from that of the maritime Hurricane test. At Emu Field the earth was heated to such high temperatures it became glass.

The second test was on 27 October. Between the two tests, Menzies defended Australia's role in the ongoing atomic test series during his regular weekly radio broadcast, *Man to Man*:

There is tremendous public interest in Atomic Bombs … Unfortunately there are scare stories, wild allegations, and, between you and me, a good deal of nonsense … But we must face the facts. And they are that the threat to the world's peace does not come from the Americans or the British, but from aggressive Communist-Imperialism. In this dreadful state of affairs, superiority in atomic weapons is vital. To that superiority Australia must contribute as best she can.

Howard Beale described Beadell as 'a man of iron endurance, and (like Kipling's elephant child) of infinite resource and sagacity'. Beadell later wrote about Old Luke, a member of his surveying team:

Old Luke had a little joke waiting at this stage for the reporters. 'Look', he shouted pointing at the atomic cloud, 'do you see it?' Everyone whipped around to direct their attention to the cloud. 'A perfect portrait of a myall blackfeller written with atomic dust; the new and old have come together today'. He was so enthusiastically serious that one by one they agreed that there was no doubt about it. Sure enough the newspapers printed the huge headlines: 'Myall black man written by atomic dust in sky over Emu'. Good old Luke.

Prominent British *Daily Express* journalist Chapman Pincher also witnessed Totem and contributed an article to the *Sydney Morning Herald*. Pincher provided a vivid word picture of Totem 1, overcome with awe at what he had witnessed:

> Peering through welders' safety goggles, I watched [the explosion] swell into a tremendous fireball – a miniature manmade sun which rose away from the red sand like a giant balloon. A minute later I was shaken by a terrific shock wave – a hot blast that sent a double thunder clap rumbling around the desert for 30 seconds. As the fireball expanded it gave off a second burst of light more brilliant than the sun.

He concluded, 'It is clear already that Britain's bomb, designed and built without outside assistance, is a winner'.

The Totem tests caused many problems. The planning was rushed and the site was endlessly difficult. The British had not tested their own bomb on a mainland desert site before, and they were unfamiliar with the weather conditions. There was also the essentially uncontrolled presence in the general area of Aboriginal people, notwithstanding the work of the native patrol officer Walter MacDougall. The two tests were destined for controversy, particularly because of the infamous 'black mist' that is said to have blinded an Aboriginal boy, Yami Lester, killed others and caused significant, long-lasting distress to all the local Aboriginal people (discussed in more detail in chapter 7). Penney later conceded when questioned by the Royal Commission that Totem 1 took place in conditions that were unsafe for all concerned.

The Totem bombs were part of a push by the British to build up the country's atomic weapon stockpile quickly. The Americans and the Soviets were already well ahead in the game, and the British couldn't be finicky about the purity of the design and execution. They had to make bombs with whatever material came to hand. The ideal material for nuclear weapons is plutonium-239, but it is expensive to produce, and British reactors were not able to keep up with the demand from the AWRE. So the Totem bombs made use of impure plutonium. Plutonium-240 is quicker and cheaper to produce, and this potentially enabled faster production of the hundreds of bombs Britain wanted. Totem was designed to test whether plutonium-240 was suitable to power nuclear bombs. Lorna Arnold, the British chronicler of the nuclear trials, called the tests 'a technical success', despite the harm that Totem 1 (in particular) caused. The tests advanced understanding of plutonium-240 for cheaper weapons manufacture, including the proportion that would produce a viable chain reaction. The British also learned about fallout patterns – effectively by experimenting on the local Indigenous population.

The RAAF deployed 10 Lincoln bombers to the Totem series to take samples from the atomic clouds. The huge military aircraft, based at Woomera and Richmond (near Sydney), landed on the Emu claypan in all weather. Also, Bristol freighter aircraft transported equipment to the site. The US Air Force sent two B29 Superfortress four-engine turboprop bombers that operated out of Richmond RAAF base, 1900 kilometres away, to collect air samples from the clouds. The American air crew who flew into the atomic cloud were provided with dosimeters, and their aircraft were fitted with radiation detection equipment. Dosimeter is the collective name for various kinds of radiation detection devices, including film badges that could be worn on the uniforms of personnel in contaminated areas.

Film badges had a plastic holder containing a piece of film similar to a dental x-ray film. Radiation exposed the film, which was later developed to determine the radiation dose received. The Americans were strict about the use of dosimeters.

This was not the case for the RAAF crew operating under British orders, or the RAF crew. The British view was that the risk of exposure of radiation caused by flying through the cloud was 'negligible', so no detection gear or personal monitoring devices were issued. Penney said in his Royal Commission statement, 'The fact that the crew of an RAF Canberra received significant doses of radiation as a result of their early passage through the [Totem 1] cloud was reported to me. I did not regard it as very serious as a once in a lifetime dose'.

In fact, only the more rigid American safety procedures drew any attention to the risk that pilots and other air force personnel faced. The Americans were stricter about safety than the British ever were throughout the time they tested their bombs in Australia. As writer Joan Smith said in her book on Maralinga, 'The Americans … knew exactly what they were doing'. American Geiger counters were run over the Australian planes as well as their own aircraft. Every time they did this the planes were 'hot' – that is, they were contaminated with radiation. The British Canberra aircraft that sampled the Totem 1 cloud, and which had been sealed before take-off, was too contaminated to be used again for Totem 2.

Nuclear armament doesn't involve just a clever design of a big weapon made of uranium or plutonium. Subsidiary matters such as triggering the devices or predicting and ameliorating the worst-case scenarios must be dealt with too. The British authorities were keen to supplement the program of major bomb trials with smaller scale experiments on a range of associated issues. They started to broaden the scope of the test program with the advent of the early minor trials at Emu Field, codenamed Kittens. These experiments were designed to test aspects of the design of bomb triggering devices known as initiators. Despite the obvious inadequacy of the Emu

Field site, five Kittens experiments were held there in 1953 before the British departed. As the British prepared to leave this difficult site, plans for what to do next were afoot. Australia was preparing for the long haul.

As future plans were drawn up, Australia's pallid attempts to become a more senior partner in the tests ran to offering more funding than the British had even requested. A top-secret 1954 Cabinet briefing document outlined this strategy:

> Although U.K. had intimated that she was prepared to meet the full costs, Australia proposed that the principles of apportioning the expenses of the trial should be agreed whereby the cost of Australian personnel engaged on the preparation of the site, and of materials and equipment which could be recovered after the tests, should fall to Australia's account. This basis was accepted by U.K., and the approximate costs of the trials in Australia amounted to:- United Kingdom £771 000; Australia 144 000. The U.K. share ... does not, of course, include the cost of development of the bomb, which probably amounted to some millions of pounds. Included in the U.K. share in Australia is a considerable quantity of stores and equipment still at Emu Field and Woomera, which could be used for future tests, or sold and a credit passed to the U.K. A cheque for £600 000 sterling was received in June [1954] from the U.K. Government in payment of her share, but adjustments may be necessary before the costs are finalised.

Beale said that Australia should aim to be more than a mere 'hewer of wood and drawer of water' for the British at the new Maralinga site. He recommended extensive financial, material and manpower contributions so Australia would be considered a true partner, after canvassing four other less costly scenarios. Beale's preferred option was projected to cost a huge £200 000 per year.

> This alternative will ensure the best return to Australia for any
> expenditure provided and comes more adequately than any
> of the other [cheaper] alternatives within the definition of a
> joint project because it means that we would have a definite
> responsibility in the scientific trials and would share in the
> knowledge gained therefrom.

In the end, despite the massive expenditure, Australia was never more than the wood hewer and water drawer. In fact, that description could not be more apt, with its connotation of lowly hard labour and exclusion from decision-making.

Harder negotiators on the Australian side might have made a difference, since Australia was not without bargaining chips, but they were nowhere to be seen. Instead, Australia just worked out how to pay the exorbitant costs associated with its menial role. After some discussion back and forth between senior government officials, they decided Australia's initial contribution should be £400 000 to £500 000, continuing at an annual rate of £150 000. Beale recommended a special government appropriation to cover the cost, rather than raid the Defence budget. The means of raising such a significant amount of money was up to Treasury – it would not come out of existing departmental or service budgets.

After Totem, the British test program in Australia stopped for two and a half years. The decision to create a permanent nuclear test site in Australia at Maralinga took a lot of planning, expense and time. While the Australian Government was preoccupied with the logistical matters associated with building a new desert township and associated bomb testing range, the British were getting impatient with the fact that the permanent site would not be ready until the second half of 1956.

During that time, AWRE scientists shifted their focus to the prospect of a British H-bomb. This fusion weapon, already in the arsenals of both the US and the USSR, had a much larger yield – in the megatonne rather than the kilotonne range, a megatonne being 1000 kilotonnes. The prospect of an international treaty to limit nuclear weapons testing gave British plans for a thermonuclear bomb greater urgency. They needed to test a megatonne weapon before a test ban took effect. Penney made the running on the British H-bomb test, in the Pacific Ocean at Christmas Island. It was given the codename Operation Grapple.

When negotiating the Maralinga agreement, the Australians had explicitly forbidden testing an H-bomb on Australian soil. The political ramifications of the much bigger H-bomb were beyond what even a compliant government felt able to deal with. However, while they were forbidden from testing the H-bomb itself, the British needed to test a highly efficient and more advanced device that could operate as a trigger for it. In a top-secret telegram prepared by the Ministry of Defence, the British made their intentions clear to the Menzies government:

> You know well the importance we attach to the speediest development of efficient nuclear weapons. You know also ... how much we appreciate the immense help given to us by Australia in having our previous nuclear trials take place on Australian territory and in agreeing to the establishment of a permanent proving ground at Maralinga. We have however still one more request to make of you. Maralinga will not be ready until September 1956. Our scientists on the other hand will be ready to make tests in April 1956 which are very urgently needed in the course of our development of more efficient weapons by the inclusion of light elements as a boost. We are most anxious not to lose the six months that would be involved in waiting for the Maralinga range to be ready.

And so, for Operation Mosaic (originally named Operation Giraffe), the British returned to the remote Monte Bello Islands for two tests, one of which became particularly infamous.

Unlike Hurricane, which was a maritime operation to test base surge caused by an atomic weapon, the Mosaic bombs at Monte Bello were detonated from land-based steel towers. The fallout was expected to be much less than from the radioactive water spout of Hurricane – Menzies was told that it be would less than one-fifth of the first British test's fallout. That was another of the false predictions that riddled the British nuclear weapons test program.

HMAS *Warrego* and *Karangi* were pressed back into service to set up the firing sites during October and November 1955. In addition, HMAS *Fremantle* and *Junee* assisted by providing transport and logistics during the first Mosaic test. The RAAF pitched in with a variety of support services, such as security, transportation and signalling. These aircraft deployed from RAAF base Pearce in Western Australia later travelled across the continent to RAAF base Amberley, near Brisbane, for decontamination. RAF aircraft also performed a multitude of tasks an atomic weapons test requires, such as surveys of ground contamination and cloud tracking.

Although there were no nuclear tests in Australia during 1954 and 1955, Mosaic was still a rush job. The experience of the 1952 Hurricane test gave the British a good idea of the challenges of the site, but the Monte Bellos were damnably remote, and the two Mosaic tests were entirely different propositions from Hurricane. From conception to first test, the British had only about 15 months to plan. For such a step-up in technology, this was recklessly fast. A lot of equipment, and of course the nuclear devices themselves, had to make the trip from the UK. On this occasion, to save time, the British took the risk of travelling via the Suez Canal, chopping weeks off the journey.

Mosaic G1 was detonated on Trimouille Island just before midday on 16 May 1956. It was controlled from a Royal Navy ship anchored in a lagoon about 25 kilometres away. It had an approximate yield of 15 kilotonnes, 10 kilotonnes less than Hurricane. The mushroom cloud rose higher than predicted, up to 6400 metres instead of 5200 metres.

The main event came a few weeks later. Mosaic G2 remains the biggest atomic device ever exploded on Australian soil. This device was also tower-mounted, this time on Alpha Island. Its yield was disputed for years and was noted by the Royal Commission as being 60 kilotonnes. Many commentators and scholars now accept it was 98 kilotonnes (although some continue to maintain that it did not exceed 60 kilotonnes, and the British have not publicly released yield data). Its cloud rose 14 000 metres, instead of the predicted 11 000 metres, and proceeded to spread east across the entire continent, causing considerable consternation in the media and among the public. Between G1 and G2, Howard Beale, who had apparently misinterpreted information from the British authorities, claimed publicly that the second Mosaic test would be smaller than the first. This assertion was so far from the truth that the statement contributed to public panic when G2 was detonated. Many people believed that something had gone terribly wrong and the massive explosion was some kind of nuclear accident. As the minister responsible, Beale had to scramble to dispel public fears.

In his autobiography Beale recounted an anecdote from his media trip to Maralinga in June 1956. (Beale's autobiography noted this trip took place in July 1956 to the Woomera weapons test range, rather than Maralinga, but since the second Mosaic test occurred on 19 June 1956, the same day that journalists arrived at Maralinga, it seems likely this was one of the several inaccuracies in his autobiographical account.) Maralinga was being highlighted at the time in media publicity and Mosaic was played down. While the papers reported on Mosaic, they were not at Mt Potter to witness it, as they had been for Hurricane. At Maralinga, visiting media representatives were alerted to allegations of what they believed were unacceptably high levels of radiation heading to the mainland from Monte Bello following G2. A rumour swept through the dining room during the evening meal, leading to a mass exodus of the editors and journalists to call their offices. The rumour suggested that a miner in the Pilbara town of Marble Bar (roughly 1000 kilometres to the northwest of Maralinga, but close to the coastline near the Monte Bello

Islands) had detected significant radiation on his Geiger counters. This mobilised the journalists and gave Beale some uncomfortable hours as he attempted to kill the rumour and restore order to the restive media contingent. Given the poor state of communications to the Monte Bello site, this took some time.

Beale, after finally receiving the reassuring words he needed from the AWTSC chair Leslie Martin, who was at Monte Bello, issued a dramatic media statement at midnight saying that most of the radioactive fallout had been deposited in the sea, although some was drifting at very high altitude and posed no risk. His later account stated:

> I learnt again from this that newspaper editors may be gracious guests, but when it comes to a sensational news story they are newsmen first and foremost. Nevertheless they co-operated in an awkward situation, and faithfully kept their word to me by damping the story down as far as they could.

That media representatives damped down the story to assist Beale is a revealing comment. It probably speaks to the kind of relationship Beale had developed with the media and the closeness to official sources that was maintained by media practitioners at the time. One story published by the Sydney *Sun* was headlined 'Threat to 3 towns'. It began, 'All Australia is anxiously watching a radioactive cloud – result of the atomic blast on the Monte Bello islands on Tuesday'. Beale was quoted citing Martin: 'Prof. Martin ... has reported to me that conditions of firing were ideal and there was absolutely no danger to the mainland. The path of the cloud was followed by plane, and last night the cloud was over the sea, 100 miles off the north-west coast'.

The story also quoted the deputy prime minister Arthur Fadden, who defended Beale and the scientists over the incident in the face of attacks from the leader of the Opposition Doc Evatt. In federal parliament Evatt indicated that the 'dogmatic statements and assurances given over and over again by the Minister for Supply' were insufficient to meet public concern. Fadden assured parliament there would be a full inquiry into the incident. In the end, there was not.

The Age in Melbourne applauded Beale, saying he 'acted promptly to allay misgivings about the cloud drift. Within a few minutes of the Marble Bar reports of heavy radioactive fall-out he readily stated all he knew, and made contact with scientists, whose assurances were soon forthcoming'. The Melbourne *Herald* had some advice for Beale and his colleagues after the Mosaic incident:

> There is a simple way in which the authorities controlling atomic weapon tests can keep the public informed and reassured about their checks on the risk in radio-activity after an explosion. Publication of regular reports by the safety committee, giving the measure of fall-out and the position of the atomic cloud, would prevent needless worry.

As Maralinga was about to begin operations, and people throughout mainland Australia were likely to be affected, this was reasonable advice. However, it was not heeded to the degree warranted by the danger of the tests.

After G2, the British were finished with Monte Bello and departed, leaving an unholy mess in a place that was effectively out of sight, out of mind. While the islands were still subject to the Defence (Special Undertakings) Act, no-one policed its exclusionary provisions. Many 'salvagers' on boats defied the warning signs and visited Trimouille and Alpha islands, intent on taking away the huge quantity of scrap metal and other waste that was left lying around the sand dunes. They did so at their own risk.

Both Monte Bello and Emu Field have faded into the background. Certainly they did not experience the same levels of plutonium contamination as Maralinga, which has become shorthand for all British nuclear tests in Australia. This obscures the fact that the nuclear weapons tested at Monte Bello and Emu showed that the British could play the nuclear arms game with the superpowers, even though they were a few years behind.

Both pre-Maralinga sites were important in the history of the British nuclear tests. The British gained much knowledge and established their atomic club credentials, at a high cost to the local people and the environment. The British, still hostage to the McMahon Act, used Monte Bello and Emu Field as staging posts. Now they had a new site, negotiated as an indefinite arrangement. The AWRE was Maralinga-bound, with high hopes that their bomb would pass its final tests and fulfil Penney's aspirations when he designed it. Soon Britain would have an operational atomic weapon that rivalled those of the US and the USSR.

4

Mushroom clouds
at Maralinga

It will be the Los Alamos of the British Commonwealth.

Howard Beale, Australian minister for Supply, 1955.

*It is marvellous how fickle the public mind is in these matters and, no
matter what the project is, provided it can be pushed ahead without
anything untoward happening, the people's minds soon become inured,
and they accept it as one of the ordinary happenings of life.*

Frank O'Connor, secretary of the Australian Department of Supply,
after the first major atomic bomb tests at Maralinga, 1956.

*I had thought Woomera to be rather desolate –
it was Piccadilly Circus compared to Maralinga.*

Major Dan Buckley, British Army, statement
to the Royal Commission, 1984.

In the mid-1950s, as the Cold War intensified, nuclear bombs
became the weapons of choice. Britain, committed to building
its atomic arsenal, had already established its credentials in Australia. Australia at that time was preoccupied with preparing for the

Melbourne Olympic Games in November 1956, a moment of great national pride and considerable distraction for a sporting nation. Meanwhile, in secret but with the confidence that came from the success of their Australian test program to date, the British were full of plans. A 'permanent' site was the next logical step, but the significant logistical difficulties ruled Emu out. Supply Minister Howard Beale told Menzies, 'Emu Field seems to be out of the question, mainly through shortage of water and difficulty of access'. A search had begun even before the Totem series in October 1953. William Penney wanted a remote location that would be suitable for both airburst tests, of atomic devices dropped from an aircraft, and ground tests, of devices detonated on or near the ground.

Again, Len Beadell played a central role. His amazing bush skills, surveying expertise and seemingly instinctive knowledge of the requirements made him best placed to look out over the land and say 'here'. He had found Emu. Now Penney turned to him once more. Beadell later recalled they wanted somewhere closer to the Nullarbor so they could use the train line, 'so I went on a 500 mile expedition, discovered a new site altogether which we called Maralinga. That was the same thing all over again – the village site, the connecting roads, the weapons area and the airstrip'. Maralinga was the final destination for the British nuclear tests in Australia. Beadell described the moment when he found the site with his small team of bush bashers: 'We all knew immediately that this was going to be the place. The saltbush undulations rolled away as far as we could see, even through our binoculars ... We solemnly wrung each other's hands and just gazed about us in all directions for half an hour'.

The site was within the Great Victoria Desert, to the north of the Nullarbor Plain. The Beadell expedition also found remarkable evidence of Aboriginal civilisation – what Chief Scientist Alan Butement, with Beadell at the time, described as the 'Aboriginal Stone Henge'. This arrangement of numerous piles of large and smaller stones and slivers of shale seemed to form an enormous arrowhead, positioned on a vast claypan between Emu and Maralinga. But, noted Butement, 'there was not time to make a detailed study of the area', and that was the end of that. The commissioning of the Maralinga site proceeded without any concern for this priceless relic of an ancient civilisation. The process included erecting survey beacons by December 1954 and choosing the locations for a permanent 3000-metre bitumen airstrip, and a road to Watson, a small railway siding settlement to the southwest. The Australian Services Task Force and the Kwinana Construction Company set out the engineering works. Kwinana was an Australian-based company wholly owned by UK firms that had not long before built an oil refinery in Western Australia. In no time, the preparations for a nuclear test site began, including building quarters for the thousands of men who would live there. Bristol freighter aircraft arrived from Britain, bringing with them the means to build a weapons testing range from scratch.

Although remote, the site, originally known as X300, was more amenable than Emu Field, with better access, more reliable water supplies and enough flat land to construct an airstrip, a railway siding and a village, built in a pleasant, heavily wooded area. Penney was overjoyed. He consulted with Butement, but Penney was the one to be convinced. And he was. Maralinga soon became one of the few places in the world where nuclear bombs were detonated.

The red desert site was officially named Maralinga in November 1953, a month after Operation Totem, and preparations began immediately to test the local meteorological conditions for their potential effect on fallout. A formal agreement to carry out atomic tests at Maralinga was signed by the two governments on 7 March 1956, following talks in London in 1955 between Menzies and Churchill's successor as prime minister, Anthony Eden. The Memorandum of Arrangements indicated that Maralinga would be available to the British 'for a period of 10 years which may be extended by mutual agreement', and the area would be rent-free. The agreement specified that no hydrogen (thermonuclear, fusion) weapons would be tested there, and that each test to be carried out would be separately agreed by the Australian Government, under the veto of its AWTSC. The document also provided for data from the tests to be shared with the Australians. The British did not often do this, however, which increasingly became a point of contention for the Australian Government.

A top-secret Cabinet minute from a meeting of the Prime Minister's Committee dated 16 August 1954 recorded the acceptance of the British request to commandeer the new site. The minutes noted that the committee agreed 'in principle to the establishment of a permanent testing ground and to co-operate with the United Kingdom in the proposed new series of tests'. The meeting also agreed to direct officials from the Treasury, Defence, Supply and Prime Minister's departments to report on the nature of the Australian contribution to, and participation in, the tests. 'These officials would need to consider in particular the ability of the Service Departments [navy, army and air force] to provide servicemen for construction and other purposes associated with the tests.' The meeting decided against sending an Australian technical team to the UK for a briefing on the series, preferring instead that the UK send a team to Australia, presumably for reasons of cost. They agreed to co-operate with the UK on initiator tests known as Kittens, scheduled for early 1955 as the first tests at Maralinga. These followed the original Kittens tests at Emu Field in 1953.

Ten days after the meeting of the Prime Minister's Committee, an interdepartmental committee meeting considered what the government had agreed. It was chaired by Frank O'Connor, secretary of the Department of Supply, and attended by high-level officials including Professor Leslie Martin, the academic physicist and defence adviser soon to head up the AWTSC. A report from the armed services departments on their respective levels of commitment to the project did not make encouraging reading. The report estimated the construction would need a workforce of between 225 and 250 personnel. The navy was unable to make any personnel available, and the army would commit personnel only if the government determined Maralinga to be a higher priority than any other 'cold war task'. Only the air force was prepared to put boots on the ground, offering a token 50 personnel for the construction task.

All three services were under multiple pressures. The Korean War in the early 1950s had sapped their resources, and these were further drained by an ongoing commitment to the South-East Asia Treaty Organisation. The interdepartmental committee referred the matter of diverting defence resources to the Maralinga project back to the Defence Department, since this 'was a task beyond the competency of the committee'.

The British were briefed on these concerns. A secret letter from RR Powell in the British Ministry of Defence to JM Wilson in the British Ministry of Supply indicated that Frederick Shedden, the secretary of the Australian Department of Defence, had 'strongly advised' them not to push Menzies too hard on this: 'Any suggestion that we regarded Maralinga as more important than Malaya might upset the Australian agreement to accept commitments in South East Asia'. Australia was somewhat torn, wanting to co-operate as a good former colony while attempting to deal with existing postwar imperatives.

This document also recorded the reluctance of the Department of Works to divert resources to Maralinga to carry out construction using civilian labour. 'It was suggested that such means of carrying out the task should be attempted only in the event of its being found impracticable to devise a plan for the utilisation of Service labour.' No-one in the services really wanted a bar of the huge work involved in establishing a massive new military facility in the Australian desert. The meeting noted that the financial contribution to the cost of building Maralinga would have to come out of the Defence budget: 'As there was no margin within the Department of Supply's allotment to provide funds for this project, the committee proposes to advise that it is unable to make any proposal to Cabinet as to what financial contribution, if any, Australia should make'.

This committee discussed the need for Australia to get some scientific and technological benefit out of the Maralinga project. It tentatively suggested that 'consideration be given to offering the services of a small scientific unit to assist in a defined operation, e.g. measurements, which would in consequence involve full indoctrination in atomic science for those Australian scientists taking part'. At this time, Australia was considering setting up the technology to create plutonium out of its uranium. If this development occurred (it never did), 'we would then be in a much stronger position to claim a right for Australian scientists to participate in the work'. Australia had not been party to the results of operations Hurricane and Totem, and some frustration came through in the minutes' suggestion that, 'as atomic weapons would be vital to Australia's defence, a firm request should be made to the United Kingdom for information on the results of future tests for strategic planning purposes'. Read in their context, these remarks seem forlorn at the very least, if not outright deluded.

The Australian Government assigned Howard Beale and his Department of Supply to oversee the development of the site by Kwinana Constructions, and the many administrative tasks associated with Maralinga. Beale was an enthusiastic servant of the project who exuded positivity from the start. 'The country itself may be described as desert, but it is deserted rather than desert and is far from being a dreary waste of baking sand', said his first memorandum on the project. This document described an Arcadian experience for the new workforce: 'The prospect is tree-studded, park-like. New buildings will be shaded by the native timber, which will not be cut except in case of absolute necessity'. The experience of living there would be similar to conditions at the larger and more established British weapons testing base at Woomera to the east. Certainly the facilities were more luxurious than many military grunts might have experienced, although the hot desert conditions were destined to defeat many a Pom.

Creating Maralinga from nothing was a huge task. A village,

airstrip, roads and other facilities rapidly appeared in the red desert landscape. The strange men-only village was in the southern part of the site, while the test area stretched northwards in a funnel shape. All the operational areas were well to the north, a series of 'forward areas' centred on colourfully named test sites. These names are evocative, if mostly inexplicable: Kuli, Biak, Tadje, Wewak, Dobo, Naya, Breakaway, Marcoo, One Tree, Kite and the most infamous, Taranaki. The 3.5-kilometre-long airstrip was a few kilometres to the east of the village. The roads to the north took on a Big Apple hue with Second Avenue, Fifth Avenue, Tenth Avenue, East Street and West Street, and they created a large grid across the massive swathes of Maralinga test site land. The village street names reminded many occupants of home: London Road, Durham Crescent, Belfast Street.

The site was managed by the Maralinga Board of Management, a joint UK–Australian organisation chaired by Frank O'Connor, with a senior British public servant from the UK Ministry of Supply Staff Australia as his deputy. The range commander, Australian Army lieutenant-colonel Richard (Dick) Durance, was directly responsible through the Maralinga Board of Management to a huge array of interested parties, including the joint chiefs of staff of the Australian military and the heads of both the Australian Department of Supply and the UK Ministry of Supply. He was also responsible to Penney, or his delegate. Radiological safety was the specific responsibility of the AWRE, which based a senior health physics officer on site. Australia placed its own health physics representative on site too, namely Harry Turner, seconded from the Australian Atomic Energy Commission. Turner was required to collect data about the radiological safety of the site and oversee safety matters at the range, although in the early 1960s he was specifically excluded from the Vixen B tests, the most dangerous experiments undertaken. Access to Vixen B was granted only to British personnel.

During the years the Maralinga site was active, a total of about 35 000 military personnel spent time there. Most (about 25 000) were from the UK, bolstered by a smaller contingent of Australians and occasional attachments of military personnel from Canada, the US and New Zealand. All were male and physically fit and most were young. A significant number of the British and Australian personnel were doing National Service. The village built to accommodate people working at the site included a number of facilities intended to encourage camaraderie, such as a swimming pool, playing fields and a theatre. By the range commander's account, it was a place of high morale: 'Considering the isolation of the area, the very trying climatic conditions for part of the year, and the diversity of the groups that make up the population, morale has been, and continues to be, remarkably high. Much of this is due to the good ration scale, basic amenities provided, and the financial gain made by the majority serving in the area'.

Maralinga had been used for the 1955 Kittens tests and other non-nuclear tests, but the first major trial was Operation Buffalo, the longest series of major trials held in Australia. By this time, Penney had handed over day-to-day control of the Australian test program to his deputy William Cook, but for Buffalo he took the helm again himself. The series, held in September and October 1956, consisted of four atomic bombs, detonated in three different ways. Buffalo 1 and 4 were detonated from towers on the ground, while Buffalo 3 was dropped from an aircraft. Buffalo 2 was exploded at ground level, the only ground-level detonation in Australia, with its concomitant risks. The other major trial at Maralinga was Operation Antler in September and October 1957 which involved three bomb firings (two on towers and one from tethered balloons). After that an international moratorium on tests, and a revived relationship between the UK and the US, put an end to mushroom clouds tests in Australia.

Nothing says 'Maralinga' more than the startling image of a billowing atomic cloud rising from the desert plain. Buffalo 1 sent up the first mushroom cloud over Maralinga. While mushroom clouds are particularly associated with atomic weaponry, any large explosion will produce the same effect, though most non-nuclear explosions don't come close to the power of an atomic bomb. Natural events like volcanic eruptions can also cause these distinctive clouds, although, of course, non-nuclear explosions tend not to be radioactive. In contrast, the detonation of an atomic device instantly creates a burst of intense gamma and neutron radiation. The cloud of an atomic explosion is filled with radioactive particles, the products of the fission process when the atoms are wrenched apart. These particles roil and swirl with the ground debris, all sucked upwards as the cloud interacts with the atmosphere. The blast also sends out a pressure or shock wave, which can cause damage to anyone or anything in the vicinity.

The height of a mushroom cloud is governed by the energy of the explosion and the atmospheric conditions at the time of the detonation. The mushroom shape forms about 10 minutes after detonation and may last for up to an hour. However, even when it has disappeared, radioactive particles remain in the atmosphere, blown by the wind and depositing fallout along the way. Fallout literally falls out of the sky. The fallout from the Maralinga mushroom clouds spread far and wide, depending upon the winds, the explosive yield and the method of detonation. The airburst test at Maralinga, Buffalo 3, produced less fallout because it drew up less material from the ground into the cloud. The yield from the Maralinga tests was far less than that from Mosaic G2, but it was still substantial, and it travelled the airways north and east.

Penney arrived at the site on 24 August 1956 and set the date for the first major test around 12 September, with the whole site to go on stand-by the day before. Rehearsals and other necessary activities got underway in earnest. Little did anyone know how protracted the wait would be. Day after day, the meteorological conditions were unsuitable. Day after day, scientists, journalists and politicians, invested in their different ways in the first Maralinga test, had to be patient. It was not easy. Questions in federal parliament and mocking stories in newspapers created uncertainty about the suitability of the site and the safety of nuclear weapons tests.

On 22 September the weather appeared to have stabilised, but an airman had 'gone missing' near the forward area. A search was organised. At 11 pm he wandered, footsore and dehydrated, into Eleven Mile camp, well to the south of the forward area, having walked 32 kilometres through the desert. By the time he was found to be safe and well, it was too late to go ahead and the test was again postponed.

After that drama, and endless frustrating delays, the all clear was given on 27 September. Operation Buffalo got underway. The first in the series, detonated from a 30-metre tower at One Tree, had a yield of 15 kilotonnes and was the most witnessed. When the long-delayed plane load of journalists arrived they were allowed to watch the awesome spectacle that is a nuclear bomb. The parliamentarians missed out. They had arrived the day before but were sent back because there was nowhere to stay overnight.

Buffalo 1 was the plutonium warhead for the future Red Beard tactical nuclear weapon, a smaller weapon than Blue Danube. The winds were still not right on the day it was fired, suggesting Penney felt that he couldn't wait any longer. Certainly the Royal Commission found that the decision was probably made out of a desire to get the thing done, rather than to adhere strictly to the agreed firing conditions. As with the earlier major trials, the mushroom cloud rose higher than predicted – well over 11 000 metres instead of 8500 metres – in a huge, classic mushroom shape. The unexpected height of the cloud played havoc with fallout predictions and appeared to contravene the conditions for safe firing that had been agreed with the Australian Government, a serious problem. An RAF Canberra aircraft flew through the cloud to gather samples. Radiation experts quickly began surveying the contamination. The radioactive cloud headed due east.

Despite some qualms about the extent of fallout from the first blast, Operation Buffalo proceeded. Buffalo 2, detonated at 4.30 pm on 4 October at Marcoo, was the only British bomb test detonated at ground level and therefore the only weapons test to create a true bomb crater. The crater was 44 metres wide and 21 metres deep.

John Moroney described Buffalo 2 as a 'nuclear landmine'. It tested the Blue Danube device that Penney had designed so many years before. It was a much smaller device than Buffalo 1 – only 1.5 kilotonnes – and had a far less spectacular cloud that rose to less than half the height of Buffalo 1. Like Buffalo 1, though, its cloud headed due east towards the northern New South Wales coast. The device was detonated despite the fact that rain had been forecast within 800 kilometres and actually fell about 160 kilometres from ground zero. Rain in the aftermath of a test brings fallout with it; any rain in the contaminated area will likewise be contaminated. Again, the agreed firing conditions were transgressed. These now looked to be optional rather than firm requirements.

Buffalo 3 was an airdrop with an expected yield of 3 kilotonnes. The winds were still difficult, and the time for detonation was brought forwards slightly. The device, also of Blue Danube design, was dropped from a Valiant bomber at 3.27 pm on 11 October and exploded 150 metres above the Maralinga plain near the Kite test site, dropping its radioactive material around the restricted area, including near to Maralinga village. As planned, the fireball did not actually reach the ground, and the cloud rose to 4500 metres. Unexpectedly, demonstrating how primitive the forecasting methods were at that time, the radioactive cloud headed to the south, drifting over Adelaide.

Buffalo 4, another Red Beard test, exploded from a tower at Breakaway, was a bigger bomb, expected to yield 16 kilotonnes, though its actual yield was 10 kilotonnes. It was fired in the dead of night, at 12.05 am on 22 October. It would have been fired earlier, except 21 October was a Sunday, and the Australian Government banned Sunday tests for religious reasons. Five minutes after midnight made everything okay, apparently. The sight of a midnight atomic fireball must have been eerie. Buffalo 4 had a high cloud, in excess of 9000 metres, but those on the ground could not see it well because of low stratus cloud. The radioactive cloud swung north and headed towards Darwin. In the tradition of previous Buffalo shots,

Buffalo 4 violated the firing conditions. Even as it was being fired, the test authorities knew that it would cause fallout in inhabited areas more than 160 kilometres from the Maralinga range, a clear violation of the agreement. In fact, it sprinkled radioactive fallout on an arc between Newcastle on the New South Wales coast and Darwin in the Northern Territory.

Buffalo had many controversial elements. In particular, this test series is notorious for the Indoctrinee Force, often referred to later as the Maralinga guinea pigs. This group, largely commissioned officers, was deliberately positioned in the forward area during the Buffalo major trials so they could witness and experience the effects of nuclear weapons close-up – less than 9 kilometres from ground zero. There were 283 men in the Indoctrinee Force. Most were from the UK – 172 officers and six civilians – while Australia contributed 100 (mostly army officers but 25 from lower ranks and one civilian) and New Zealand contributed five officers. They were under the direct command of Australian captain JH Skipper but under the general direction of the British scientist Drake Seager, who reported to Penney.

The members of the Indoctrinee Force were special. They were housed separately from the other Maralinga denizens. They stayed at Eleven Mile camp, which was 18 kilometres from Watson and about 64 kilometres south of the forward area. They received end-less briefings, lectures and range tours. Because the first Buffalo shot was delayed for 15 days, the Indoctrinees also assisted the scientists in preparing and laying out the various objects that were to be subjected to the nuclear blasts, such as guns, cars and dummies. The Indoctrinees witnessed the first two Buffalo tests, One Tree and Marcoo, up close. The world was facing the real prospect of nuclear war, and the Indoctrinees were ordered to report back to their military colleagues what the future had in store. Their eyewitness accounts were expected to provide preparation for the reality of atomic warfare.

Major Peter Lowe finally arrives at Maralinga, after a long British Overseas Airways Corporation flight to Sydney via the Far East. He has 'been volunteered' for a special three-week mission in the Australian desert. His commanding officer called him, when he was working at his post in Münster, West Germany, to tell him the glad tidings. It is all very hush-hush, he was told, but there is a big show on in Australia that needs non-technical observers, and most of them are coming from the British Army. He was to join an elite group known to the AWRE as the Indoctrinee Force, or I-Force, a sinister-sounding designation, though not one the men themselves use or even know about. Major Dan Buckley, also in the British Army, soon joins Lowe. He was stationed at Woomera for the rocket tests not long ago and had no idea he would be back in Australia so soon, for an even more dramatic assignment. Buckley is young, sporty and fit, a boxer and rugby player in his spare time. Lowe and Buckley are in different Indoctrinee teams doing different things, but both form part of a major exercise at the first Maralinga major test series.

Their camp at Eleven Mile is tented and primitive, rather different from those in Münster or the UK – or even Woomera. They are away from the main Maralinga contingent in the tree-lined village, too. It soon becomes clear that their mission will take longer than three weeks, as the winds are never right and interfere with the test schedule. Time hangs heavily on the officers. While they wait, they are treated to lectures and other preparations. Penney's lectures are interesting and well prepared. The same cannot be said for some of the other talks, which are boringly technical and hard to follow. But everyone enjoys hearing Sir William speak, even though his subject matter is grave. He has an amiable and egalitarian manner, and knows his stuff. Penney warns the men of the dangers of gamma rays and describes the measures that will be taken to protect them. They are to wear full protective suits and film badges. Designated members of the party will carry Geiger counters.

The Indoctrinees are frustrated by the delays to the first blast at

One Tree. After the lectures, they take the time to prepare equipment that will be subjected to the atomic blast. They are officers, and this hard physical labour is not normally the sort of thing they do, but they are bored enough to welcome the activity. Lowe helps dig in 25-pounder guns, erect radio antennae and set out field telephones. Buckley has a particular responsibility for guns and is asked to ensure a range of damage from none at all to complete annihilation. He spaces his guns out from ground zero to achieve an even coverage. He has been told that this work will have implications for British Army equipment purchasing policies – guns that seem to survive an atomic blast will require fewer spares than those that are quickly destroyed.

When the day finally arrives, after a number of false starts, the Indoctrinees are stationed at Forward Control on a hillside only 8 kilometres away from the first explosion – far closer than the main party. They are all dressed in summer uniform of shorts, shirt and long socks. A minute before the blast, as one they turn around so they are facing away from the explosion. Buckley listens to the countdown – '4, 3, 2, 1, flash turn now!' At 'flash' the sky explodes, and it feels like an oven door has been opened right next to their bare necks and knees. They can turn now, as the initial flash is over and the possibility of eye damage has lessened. Buckley spins around to see a coiling black mass, shot through with flames, rapidly reaching higher for the colder air and then gradually flattening into a mushroom shape. He and his colleagues see the blast wave rushing towards them, knocking over the vegetation in its path. Then the wave hits and the men rock on their feet.

After the blast, the Indoctrinees venture towards ground zero to see what the explosion has wrought. Lowe dons his gas mask, boots and the protective clothing they call goon suits and climbs onboard a 3-tonne truck to travel to the edge of the contaminated area. He heads a small group of eight men and holds the Geiger counter. The device gets more and more frantic as the men near the blast site. The conditions are clearly too radioactive, so the group heads back to the decontamination camp run by the Australians under the direction of Harry Turner near Roadside, a waypoint at the junction of the network of Maralinga roads. Lowe's film badge is ripped unceremoniously from his lapel by a big, tall, brusque Australian sergeant and thrown into a bucket, without its number or reading being recorded. The bucket is full of film badges, none of which will ever be seen again by those who wore them.

Meanwhile, Buckley nears his forward gun location and is puzzled to see that the arid reddish-brownness of the desert earth has been replaced with a strange whiteness, the earth transformed by extreme heat into white glass. He reaches his leading gun position and calls out his Geiger counter reading to the health physics representative. The reading is off the dial, and he gets the urgent message from his base 'Return immediately, return immediately. Report to Health Control'. He doesn't need to be told again – he hurries away from the unnatural white glass that overlies the red dust.

When the Geiger counters tell the team that it is safe to go back out, Buckley gathers together the sacrificial weapons that he placed in the forward area. He needs to test fire the guns that are still in one piece to determine what damage they have incurred. This is a laborious process that requires a strict safety protocol. One of the senior brass tries to hurry him up, suggesting that if Buckley is afraid to fire the guns himself then he will do it for him. Buckley is insulted and fires the next gun without taking his usual precautions. It blows up, nearly bursting his eardrums and leading to endless ear tests later on back home. So much for losing his temper and neglecting his safety training. Not long after his service at Maralinga, Buckley will retire from the British Army, and a few years later his health will fall to pieces. He will develop cataracts in his eyes, the blood disease haemochromatosis and severe arthritis, which will curtail his burgeoning post-service business career in the exports sector, a career that will earn him an OBE for services to UK exports.

Unlike Buckley and most of the other Indoctrinees, Lowe does a second stint in the forward area. He takes over from a sick colleague at the last minute and witnesses the second Buffalo blast, the ground shot at Marcoo, from a Centurion tank with two other Indoctrinees. This is incredibly scary, partly because he mistakenly believes it is an airdropped bomb and he fears for the accuracy of the bombardier. He does not know how far the tank is from ground zero, but at the moment of detonation he knows the tank moves about 3 metres sideways, a claim that will later be contradicted by a more senior officer. But he is there, and he knows. The massive tank moves like someone has picked it up. He watches through the periscope, which goes opaque at the moment of detonation because it is sandblasted. When he gets out of the tank, 30 minutes after the explosion, he sees the paint on the tank has blistered. Lowe is wearing ordinary light military clothing and no film badge. Later the British Army will deny that Lowe was inside the tank at Marcoo. His army record will never show that he witnessed Buffalo 2, and he will have trouble obtaining a war pension.

Lowe will later be promoted to colonel and will start to experience health problems when he is a military attaché in Washington in 1969. His severe gastric problems will be something of a mystery. In 1972, working as a military adviser to the British High Commission in Canberra, he will have an internal haemorrhage. After some diagnostic confusion, he will finally discover that he has stomach cancer and will have his entire stomach removed.

After Buffalo, the AWRE turned its attention to Antler, the next series of Maralinga tests. The second half of the 1950s was a pivotal time for atomic weapons development. Both the US and the USSR had performed airdrop tests of hydrogen bombs, weapons of huge yield, which led to growing concerns about what the tests were doing to the earth's atmosphere. The UK herself was about to test a hydrogen device in the Pacific. Coupled with this, international tensions – particularly the Suez crisis in July 1956 and the Soviet suppression of Hungary in November – added to a sense of general foreboding. With bigger, more deadly atomic weapons, and world events seemingly on a conflict trajectory, public disquiet about atomic tests increased. Intellectual movements such as the 1957 Pugwash Conference of scientists and other scholars opposed to nuclear weaponry emerged. The UK's Campaign for Nuclear Disarmament was launched in 1958 and quickly grew in size and influence. The mood was shifting.

Howard Beale was the minister who had to balance growing public fears in Australia with British test requirements, at least until his abrupt departure from Cabinet in early 1958 to become Australia's ambassador to Washington. His role was, in part, to use the techniques of public relations to maintain effective information management around the tests. In his 1977 autobiography the chapter on the atomic tests began by revealing his success at keeping the details secret. The French had called the Australians hypocrites for objecting to French tests at Mururoa Atoll in the Pacific after allowing the British tests in Australia. Beale noted that when this happened, 'many people were taken aback, not at being called hypocrites … but to learn that we had ever conducted tests at all'. Beale then told his own (rather inaccurate in places) version of the tale, saying, 'It is not one of which any Australian need be ashamed'.

Beale was often directly involved in media activities around the atomic tests, as was William Penney; on some occasions they were a double act. Strictly controlled interactions between journalists and senior test scientists or government officials were held from time to time to diffuse unauthorised journalistic inquiry. The Supply secretary Frank O'Connor rather wistfully wrote to the chief information officer for the UK Ministry of Supply, Iyer Jehu, on 9 November 1956 in relation to a Chapman Pincher story about nuclear testing activities. O'Connor described Pincher, who covered the tests in Australia for the British *Daily Express*, as a 'scoop journalist employed by a scoop newspaper and the moment he stops scooping he will be replaced by someone else'. He observed that 'philosophically, we have to recognize this is just part and parcel of the democratic set up ... My own view of the press is that it is imperative to have good relations with them, but as to whether our relationships are good or bad is a matter that is mainly in our own hands'.

When the first Buffalo shot was delayed for a couple of weeks, Beale had to deal with media speculation and hostile parliamentary questioning that suggested Maralinga might not be the ideal place for a permanent test site. The lead-up to the Buffalo series was uncomfortable for Beale, the most recognisable face of the tests in Australia, because Maralinga was a new, expensive, untried venue. A huge front page banner headline in the Sydney *Sun* on 25 September 1956, 'Latest on the bomb!', had directly beneath it in large type an actual cable from the reporter, saying, 'Hope to be back by Xmas. In meantime could you [send] further £15. Have done 6/700 word special on whether £6-million Maralinga is a white elephant'.

The story went on to explain that the *Sun* special reporter who composed the telegram had been 'waiting for a fortnight for scientists to set off an atom bomb in the first of several tests'. It quoted the reporter asking, 'Have the British and Australian governments blundered in picking Maralinga as their test site?' More people were asking this question, including the Opposition. The government, with Beale leading the charge, consistently defended Maralinga.

Beale was in the public eye throughout the Buffalo program, dealing with both media and political pressure. The Australian Labor Party (ALP) leader Doc Evatt attempted a censure motion in parliament to condemn the government's support of British atomic tests. The Sydney *Sun* reported that Beale had taken the opportunity to reiterate that the test program would continue. 'Beale said that the Governments of Australia, Britain and the US were doing their best to achieve a working agreement, which would allow abandonment of atomic tests and the nuclear arms race. "But we have no intention of stopping until we get some form of safeguard that the Free World will not be overwhelmed by an avalanche of Russian atomic arms", he said'.

A remarkable two-page memorandum showed that 1956 was a watershed year in media and public perception of the British tests. A departmental briefing document titled 'Press Reaction to Atomic Trials' set out in terse numbered points an overview of the attitude of the media to date. It noted the favourable early media treatment after Totem in October 1953, before public opinion began to turn in 1954, 'partly due to the death of a Japanese fisherman injured by radio-active fall-out from an American H-bomb explosion in the Pacific'. The memorandum also pointed to other factors changing community attitudes, including the 'apparent Soviet policy trend towards peaceful co-existence' and the international Peace Campaign 'fostered during 1954 and 1955 by people of many shades of thought'. It described how the press, 'chameleon-like, began to offset any articles showing pride in British technical advances with far more attention to the dangers, both political and physical, implicit in atomic trials', and in particular how 'the *Truth/Mirror* chain of papers, began definitely to oppose any further tests in Australia'.

The memorandum specifically noted the impact of Mosaic G2. A Gallup poll in March 1956 showed a majority of Australians 'apparently against' tests, although Western Australia and South Australia, the two states where they were held, came out in favour. So while the announcement of the permanent test site at Maralinga 'was soberly and well presented by the major national newspapers', newspapers felt compelled 'to be critical of atomic tests because of the whipped up, if unthinking, public outcry against them. The near-hysteria built of flimsy misconceptions following the second Mosaic explosion is indicative of the difficulties now to be faced from a volatile Press, public opinion and political situation'. The writer concluded by arguing for a concerted campaign 'to refute the major false issues' concerning trials and for public education through newspaper and magazine articles 'to restore the confidence and pride which only three years ago marked the ordinary Australian's attitude towards co-operation with the U.K. in this vital defence matter'. But the Australian public never again evinced the patriotic fervour that had greeted Hurricane and Totem. If Mosaic had turned the tide, Maralinga increasingly made them uneasy. The public relations machine could not fix this entirely. After Operation Antler the media were shut out, and the dangerous Vixen B minor trials that followed Antler were not reported.

By 1957, then, public opinion had hardened towards the tests generally, and the world was moving towards new treaties that would limit atomic testing. Bipartisan political support for the tests had collapsed in the wake of the second Mosaic test, and the Opposition began to ask questions in parliament about their continuation. In June 1956, after Mosaic G2, the ALP caucus voted to oppose future atomic tests in Australia. The Australian media reported that political pressure was escalating as the ALP moved to an anti–nuclear weapons policy stance.

> The next Federal Labour [sic] Government would vote no money for tests of nuclear weapons or the development of means of waging nuclear warfare, the Deputy Leader of the Federal Opposition, Mr Calwell, said today. Mr Calwell said that developments after the Monte Bello atomic explosion last week showed that if nuclear weapons tests were permitted to continue there, radio active dust, despite all precautions, might be carried across Australia.

Pressures were bearing down on the government from growing anti-nuclear sentiment in the community and increasing questions in parliament. Mosaic G2 had been a huge political issue for Menzies, and he did not want it repeated at the new test site. Also, Australia had by now grown rather tired of being kept in the dark about the details. The proposal for Antler was typically vague, with the main point being the plan to test five bombs (later amended to six, although in the event only three were tested) in tower-mounted trials. As Operation Grapple was gearing up to test a British H-bomb in the Pacific, the government wondered if the British planned to defy the terms of the Maralinga agreement and test a thermonuclear weapon. In some ways this seemed likely, since thermonuclear weapons were now the main game and Maralinga was the permanent British test site. The terms of the Maralinga agreement had not exactly proved an insurmountable obstacle to the British before.

Consequently, approval was slower than usual in coming. The Australians first received a request for the Antler series on 20 September 1956 but took until 16 May 1957 to grant approval. Despite this uncharacteristic delay, Antler took place as scheduled in September 1957. Initially the British named this series Operation Sapphire. However, in early 1957, without warning or explanation, they changed the codename to Operation Volcano. The horrified Australians rejected it outright. The name suggested violence and destruction. Antler was chosen after the Australians voiced their concerns.

Antler was connected to the H-bomb trials in the Pacific and designed to test certain components necessary for thermonuclear weaponry. Penney was not the director this time around. Instead, it was led by Charles Adams, who had been second in command at the Mosaic tests. The tests began at Tadje on 12 September 1957 with a small 1-kilotonne device (known as Pixie) detonated from a 30-metre tower. Round 2 (Indigo Hammer) was held on 21 September at Biak, again detonated from a tower. This device had a yield of 6 kilotonnes. The final weapons test in the Antler series was held on 9 October at Taranaki. The device, a trigger for a thermonuclear weapon, was suspended 300 metres aloft in the desert sky by balloons. With a yield of 25 kilotonnes, the same as Hurricane, this was the biggest device tested at Maralinga. The balloon tests could hoist atomic devices far higher than towers. This was of great interest to the AWRE, and development work had started at Aldermaston about 18 months before the test. Ernest Titterton observed the progress of the balloon systems when he visited the UK in March 1957.

The second Antler shot was observed by a grab bag of about

70 international visitors, including representatives from the Central African Federation, the Federation of Malaya (now Malaysia), the North Atlantic Treaty Organisation, the Baghdad Pact (Iraq and Iran) and two South-East Asia Treaty Organisation partners, Thailand and the Philippines. The third and final shot was observed by representatives of the international and Australian media and a range of Australian parliamentarians.

By the end of 1957, Howard Beale, the Australian face of the test program, had apparently fallen out with Menzies, and early the following year he shipped out to a post as ambassador to the US (the posting of choice for fallen Australian Cabinet ministers).

As it turned out, Antler was the last major series. The international situation had changed markedly. Britain was now a signatory to an international moratorium on atomic weapons struck in Geneva. Operation Lighthouse, scheduled for 1958 at Maralinga, never happened. No doubt this was a good thing, since Lighthouse involved more extensive human exposure experiments along the lines of the Indoctrinee Force – hundreds of men wrapped in special blankets were to huddle close to a major explosion.

There were no more major atmospheric tests at Maralinga. The seven nuclear devices tested there had sent radioactive clouds over a large portion of the Australian continent, joining the contamination already contributed by the non-Maralinga tests, particularly Mosaic G2. To this day no-one can say for sure whether this contamination caused harm to the broad Australian population. Almost certainly, though, the people closest to the bomb blasts, the military and scientific personnel at Maralinga and the Indigenous people in the vicinity, were physically affected to varying degrees.

This is not the end of the story. If the British nuclear tests had involved only the mushroom cloud tests, most of the radioactivity would be undetectable now. But there was still much to find out about the innumerable intricate details of nuclear weapons design. Despite the moratorium, the British were not to be denied the knowledge that minor trials could provide. Maralinga continued to help the British understand what the atomic age had unleashed. The minor trials that burgeoned at Maralinga after the major trials ended sometimes strayed into murky legal waters. And while much of what happened at Maralinga beyond the big, showy mushroom clouds was considerably more damaging to Australian territory, it was not uncovered for decades.

5

Vixen B and other 'minor trials'

*I'm sure in 1985 plutonium is in every corner drug store,
but in 1955, it's a little hard to come by!*

'Dr Emmett Brown', *Back to the Future*, 1985.

*In view of the known long half-life of plutonium (24 000 years),
the Vixen series of minor trials should never have been
conducted at Maralinga.*

Royal Commission into British Nuclear Tests in Australia,
Conclusions and Recommendations, 1985.

*And the more questions we asked, and the deeper we got into the
issue, the more it looked like a Pandora's box.*

Tom Uren, *Straight Left*, 1995.

In the red dust of Central Australia, sweating men in cotton shorts
are erecting a scaffold. The 2-metre-high structure is crude but
strong – narrow triangles that stand on four legs and reach a point
at the apex, where there is room to place some lead bricks. The
whole structure weighs about 60 tonnes. It has a hole at the top

to accommodate a simulated warhead. The Australian military men work quickly, craning and bolting the prefabricated structure into place, brushing away the flies and drops of sweat from their eyes as they work. Close by, the decontamination team has set up a station. It includes a cattle grid that is used for washing down the vehicles. There is also a steam cleaner and pumps, with associated water storage tanks. The ground is well covered with low scrub and some grass, with the exception of the area inside the perimeter of the Taranaki test site. That area is mainly bare sand. The temperature climbs above 40 degrees Celsius. The red sand swirls endlessly, irritating their eyes. The horizon shimmers far in the distance.

Soon the British Royal Engineers will arrive with the simulated warhead. Inside the warhead, a sphere of silvery metal is held in position by a simple bracket and metallic rods of the same substance. Placed around the silvery metal, like a suicide-bomber's explosive girdle, is a ring of TNT rods, each with an outer case of distinctive red cardboard. The silvery metal is plutonium, one of the most deadly materials known. This strange and sinister device will be exploded from the top of the metal scaffold. The scaffold is called a feather bed, for reasons that have been forgotten. The delicate mock warhead is placed in position at the apex of the steel structure and tightly secured. All is ready for the experiment.

British scientists and military personnel are in the forward area. The Australian personnel have gone well back from the Taranaki test site, mostly south to Maralinga village. Only the British can be close to the main action. Not far away, eight hydrogen-filled balloons are bobbing around at two different elevations in a circle, around the forward area as well as upwind. Each balloon bristles with measuring equipment. Six form a ring around the forward area and carry instruments to measure the outpouring of radiation from the simulated warhead when it is exploded; these are positioned at a height of just over 100 metres. Two upwind balloons, positioned at just over 300 metres, will measure weather conditions at the time of the tests. The weather determines where the contamination spreads from atmospheric tests, whether radiological or nuclear. The balloons add a surreal feeling to the outback range as they silently wait for data from the coming blast and the wind.

An array of measuring instruments designed to catch the falling radioactive debris is set out at intervals of 1600 metres around the boundary of the forward area, along East Street, 25th Avenue and West Street. A balloon team and a photographic team support the six-man scientific staff, headed by Major JT McLean. Two Holden cars, six Land Rovers and a bus are parked just outside the forward area. The forward area is generally taken to be anywhere north of Roadside, a command post at the point where the road north from Maralinga village branches into two roads. Several Land Rovers inside the forward area will be sacrificed to contamination, one of the costs associated with radioactive testing. One day they will be buried here.

It is September 1960, and the first Vixen B trial is about to begin.

Kittens, Tims, Rats, Vixen. These strange, incongruous words, no doubt small in-jokes among the nuclear insiders, labelled some of the most secret and, in the end, most damaging activities from the 11 years of British nuclear testing in Australia. Hundreds of tests fell under the heading 'minor trials', a constant since Emu Field, culminating in Vixen B. While mushroom clouds announced the major trials, the minor trials were more shadowy. And much more deadly.

Were the minor trials the most dangerous scientific experiments in Australia's history? In the case of Vixen B, the evidence is strong. These experiments were covered in unprecedented secrecy, far more so than the major trials. In part this reflected the growing unpopularity of the weapons program at Maralinga and the growing international constraints on testing. Indeed it's likely that secrecy surrounded Vixen B more for political than for military or national security reasons, and these tests may not have been possible without it. Had the general public been aware of the danger of Vixen B, the political backlash might even have swept the Menzies government from power, given its shaky standing at that time in the polls. (At the December 1961 federal elections, just over a year after Vixen B got underway, the government suffered a significant electoral downturn and was returned with a majority of just one seat.)

History has judged the minor trials harshly. They left by far the biggest portion of the radioactive contamination in Australia and were the subject of an active cover-up by the British, as *New Scientist* journalist Ian Anderson later revealed (see chapter 11). Lorna Arnold said, 'The minor trials had left more trouble behind them than the big explosions'. They could have been carried out in the UK but for politics. Noah Pearce was in the team conducting the trials. When Counsel Assisting Peter McClellan suggested to him at the Royal Commission that 'the planning foundation for your work was that radioactive contamination of Australia may be politically acceptable but not for the UK', Pearce's answer was a simple 'Yes'. This was not disclosed to the Australian people at any time during the experiments.

Might the experiments have been conducted more safely if they had been held in the UK? Probably, given the more rigid regulations and more intrusive and active media there. The experience was different, too, when the UK did similar experiments with the US in Nevada under the name Roller Coaster. Radiation scientist Peter Burns later observed, 'The Americans had a much more intensive assessment of the fallout by taking many samples. They had labs on site at Nevada so they could take soil samples and do their measurements ... they were determined to find every bit of plutonium on the ground so they did a very detailed study of what was there'. Greater rigour, far more extensive documentation and monitoring, and a proper clean-up afterwards: so different from Maralinga, where large quantities of plutonium were left lying around in the open.

The aftermath of the minor trials dominated the media stories that emerged in the late 1970s. In the mid-1980s, the McClelland

Royal Commission condemned them, which boosted the momentum of criticism. ALP deputy leader Tom Uren, who remained angry about the trials for many years, and outspoken nuclear veterans such as Avon Hudson, also raised their profile. The courts heard multiple claims from nuclear veterans, some seeking compensation for health problems caused by alleged contamination from Vixen B.

Nearly all the minor trials tested how radioactive and other toxic substances would react when burned or exploded. Different substances were used in each series, including beryllium, uranium and several isotopes of plutonium, as well as short-lived radionuclides such as polonium-210, lead-212 and scandium-46, which the Royal Commission found had decayed 'to insignificant amounts' since the time of the tests.

The first Kittens trials at Emu Field were intended to test 'initiators' – mechanisms within the atomic assembly that switch on the supply of neutrons to enable a chain reaction. The experiments concerned timing the release of neutrons used by a fission weapon to split the atom. The five Emu Field Kittens experiments used the toxic (but non-radioactive) element beryllium as well as the short-lived radioactive element polonium, releasing both into the local environment. William Penney said the Kittens experiments were undertaken in Australia rather than in the UK 'since they could be done in conditions where dispersal of the short-lived radioactive material used in the initiating of the nuclear explosion would not pose a hazard'.

Rats and Tims were held at Maralinga. The Rats experiments measured how materials were compressed under the high pressure inside a nuclear warhead when it was detonated. Tims was similar to Rats but used a different measurement method. In the Tims experiments, one of the materials measured was plutonium, so Tims left more significant contamination.

They were initially called minor trials, and this was followed by other innocuous names – in 1959 there were 'assessment tests' and from January 1960 the Maralinga Experimental Programme, often abbreviated to MEP. Justice James McClelland commented on the 'almost comical touch of camouflage in the changes of name of the minor trials', especially given the ban on nuclear tests being negotiated at the time.

Vixen B investigated questions about safety of storage and transportation of nuclear material. How would bombs and related paraphernalia behave if, for example, a plane laden with nuclear warheads crashed on take-off? The 'broken arrow' scenario – 'an unexpected event involving nuclear weapons that results in the accidental launching, firing, detonating, theft or loss of the weapon' – loomed large after several crashes of aircraft carrying atomic weaponry. Since 1950, a large (and mostly secret) array of accidents involving nuclear weaponry had occurred, sparking fears of a nuclear catastrophe caused by accident or terrorism.

When Vixen B was first planned in 1958, Britain had undertaken six years of atomic weaponry testing, and its nuclear arsenal was sufficiently advanced to go into operational deployment. Blue Danube was a tactical nuclear weapon (soon to be replaced by the smaller Red Beard), deployed to the RAF. When the Vixen B series began in 1960, Britain had rejoined the newly amenable Americans and, among other things, embarked upon a series of similar tests. The US Roller Coaster tests, which examined environmental dispersal of plutonium, had slightly narrower objectives to Vixen B, but similar methods. The Vixen B tests in Australia explicitly addressed the broken arrow scenario in addition to plutonium dispersal. Some people also suspected that Vixen B was more than just a safety test series. According to nuclear engineer Alan Parkinson, 'While they were said to be to test the safety of nuclear weapons in storage or transit, there was also an element of weapons development in these trials'. The data from Vixen B are still retained by the UK Ministry of Defence, long after the 30-year rule, meaning no-one is entirely sure exactly what they found or how it was used for weapons development.

The 12 Vixen B experiments at Taranaki between 1960 and

1963 involved blowing up plutonium-239 with conventional explosives. Nuclear warheads were strapped on feather beds 2 metres above the ground and subjected to so-called one-point safety trials. This meant detonating one point in a matrix of, say, 32 points of high explosive that must all explode in a rapid sequence to ignite an atomic blast. The plutonium was contained within the bundle of explosives. When the explosive was detonated, the plutonium was compressed and became molten. Plutonium is pyrophoric – it burns on contact with air – and when blown up it produced an aerosol of plutonium oxide particles that spread out from the Taranaki site. Narrow plumes of plutonium aerosol stretched many kilometres out in a hand-like shape over the northwest to northeast. The one-point trials were intended to show that igniting one point in the matrix would not set off the nuclear fission of an atomic blast and apparently succeeded in doing so, at least for low-yield weaponry.

This form of experiment was dangerous, as the British acknowledged in their secret correspondence. AWRE safety co-ordinator Roy Pilgrim noted in a memorandum 'the potential catastrophic nature of a mistake' and urged 'rigid adherence to the planned procedure'. The tests were not intended to produce nuclear reactions, but, in the event, as secretly predicted by the British, fission and fusion reactions did occur. Indeed, Pilgrim discussed fission openly with Ernest Titterton in a letter of October 1962, saying, 'Whereas for previous Vixen B firings the experiments were so designed that fission products could not be present in quantities sufficient to add a radiotoxic effect ... we are now seeking greater flexibility in the design of the experiments and to achieve this we need the freedom to plan in such a way that fission products may be generated'.

While the Vixen B test series used most of the plutonium-239 that contaminated Maralinga, various isotopes of plutonium were used extensively in other minor trials too. In one Tims trial, half a kilogram of weapons-grade plutonium was fired into a pad filled with salt, and six drums containing the contaminated salt were then buried at the Maralinga airport cemetery. This plutonium was mostly plutonium-239, along with some shorter lived isotopes – plutonium-240 and a tiny amount of plutonium-241. The plutonium from this test became the centre of a media controversy in 1978 when a secret Cabinet submission revealing the burial site was leaked to journalist Brian Toohey (see chapter 10). It was repatriated to Britain in 1979, the only loose plutonium from Maralinga to be recovered.

These high explosive tests ceased in 1963 when both Britain and Australia became signatories to the United Nations Partial Test Ban Treaty that outlawed atmospheric testing. The official name for the partial test ban treaty was Treaty Banning Nuclear Weapons Tests in the Atmosphere, in Outer Space and Under Water. It was signed in Moscow on 5 August 1963. An earlier moratorium on nuclear weapons testing, from 1958 to 1961, may have been knowingly subverted by the British test authorities in the case of the Vixen B tests, largely through the use of innocuous names ('minor trials', 'assessment tests'), and without telling the Australian Government exactly what the tests involved. John Moroney wrote later:

> Both [the UK and US] believed that these [one-point]
> studies were not nuclear weapons tests within the terms of
> the moratorium, but they were anxious not to be seen to be
> infringing the terms in any way. Accordingly, they performed the
> tests on reduced assemblies of the fission triggers to ensure that
> any nuclear yield was small, and conducted them under tight
> security, away from prying eyes.

As a direct result of the Vixen B tests, the feather beds, and lots of other equipment and buildings in the area, became impregnated with the most dangerous plutonium isotope, plutonium-239. The explosions created a kind of 'dirty bomb', releasing significant quantities of radioactive material into the atmosphere and subsequently onto the ground. In a submission to the Royal Commission, the ARL (now ARPANSA) 'estimated that there were between 25 000 and 50 000 plutonium-contaminated fragments in the Taranaki area, although the number might need to be doubled if missed and buried fragments were included ... The finding of this large number of plutonium-contaminated fragments was a surprise and changed the whole concept of hazard assessment of the plutonium-contaminated areas'. Over time, this calculation was revised to three million fragments. Staging the Vixen B trials cost Australia a lot of money – the cost for the first series alone was over £25 000 in labour, materials, plant hire and other expenses. It reaped three million loose fragments of plutonium.

Small particles of plutonium can be picked up readily in dust and can swirl around the landscape. Anyone in the vicinity might breathe it in. It is an insoluble particle that, if inhaled, lodges in the lungs, where it can stay throughout a person's lifetime and irradiate its surroundings, possibly causing lung cancer. The risk is precisely proportional to the dose. If a person breathes in enough plutonium-239 to receive a dose of 1 millisievert (a unit of biological absorption of ionising radiation) the risk that he or she will get lung cancer is about one in 20 000. If a person ingests or inhales enough to receive a dose of 100 millisieverts then the risk is one in 200.

Leaving such a substance lying around on the ground was reckless. The widely dispersed Vixen B plutonium was not enough to kill people immediately through radiation sickness, but it could cause cancer over longer time frames. Given the 'dusty lifestyle' of the Indigenous population in the area, this was an unacceptable risk according to accepted international guidelines on the use of radioactive substances. While the AWRE maintained it followed the protocols laid down by the International Commission on Radiological Protection at the time of Vixen B, the plutonium contamination around Taranaki shows these assurances to be unfounded. The commission protocols relevant to the British tests were established in 1950 and updated several times during the test program, most notably in 1958. By the time of Vixen B these protocols had established that there was no threshold above which exposure became danger-ous. Any exposure was dangerous.

Another risk was visitors coming to the site and picking up 'souvenirs', also unacceptable under the guidelines. To this day, no-one knows if such mantelpiece ornaments are out there – for several years during the 1970s the test range was not patrolled, and anyone visiting the area could have picked up a lump of plutonium-soaked rock or metal. The Commonwealth Police provided secu-rity services at the Maralinga site throughout the test program and remained there until 1 March 1974. In December 1976, when stories started appearing in the South Australian media about the Maralinga aftermath, the Australian Federal Police resumed surveil-lance. In between, only two civilian caretakers were on site.

In 1979, as the Maralinga story was breaking in the national media, South Australian scientists found that 19 rabbits around the Taranaki site had taken up a variety of radioisotopes in their fur, including plutonium-239, and this was cause for some con-sternation. In their report, quoted by Australian journalist Robert Milliken, a prominent chronicler of the British nuclear tests, they noted, 'It is possible for rabbits, that are notorious for their ability to excavate burrows in almost any material, to gain access to the [Taranaki test debris burial] pits by simply burrowing under the 6 inch concrete slabs … As we are discussing products that have a half life of 24 000 years, it would seem almost a statistical certainty that in some time in the future the rabbits may have access to a pit'. The pits were dug into limestone, which formed the walls of the pits, and capped with concrete.

After the well-controlled media coverage up to the mid-1950s, from 1957 onwards journalists stopped writing stories about the British atomic tests. Once Howard Beale had gone to Washing-ton and William Penney was engaged in Operation Grapple in the Pacific and, later, with the UK Atomic Energy Authority, the ongoing activities at Maralinga were not reported. However, both the British and the Australian authorities knew the Vixen B trials might attract media interest and planned for it. A sequence of correspondence in the second half of 1960 disclosed some of the official thoughts shared between the respective governments.

On 27 September 1960, Maurice Timbs, assistant secretary in the Prime Minister's Department (and from 1964 to 1973 an executive member of the Australian Atomic Energy Commission), sent Menzies a statement drafted by the British, to be used 'in the event of any public disclosure of the existence of these experiments [Vixen B] ... The intention is that it will be held in readiness and released only if there is a public disclosure that these experiments are being carried out'.

The letter had a handwritten annotation, above Menzies' initials: 'Discussed with Mr Townley [Defence minister] and approved as amended'. An attached media statement with handwritten corrections asserted that no nuclear explosions were being carried out on the Maralinga range. More detailed information about the activity at the range was crossed out, in particular a statement that the experimental program involved radioactive or nuclear materials. What remained was the following:

> The Range is being used for experiments conducted on
> behalf of the United Kingdom Energy Authority which has a
> need to explore systems of safeguards [the previous few words
> crossed out by hand] to eliminate or to minimise the hazards
> which could arise from accidents involving radio-active
> materials. The Australian Government has agreed to the use
> of Maralinga for these experiments which are carried out in
> accordance with the requirements of the Safety Committee
> established by the Australian Government and under carefully
> controlled conditions to avoid any significant radio-active
> hazard.

On 20 October, the office of the UK high commissioner in Canberra replied, unhappy with 'systems of safeguards', saying it 'may lead to difficulties and misunderstandings' because it was similar to terminology being used in negotiations for the new Geneva nuclear weapons treaty. The International Atomic Energy Agency (IAEA) was also using the term in regards to civil uses of nuclear technology. 'Thus the term "systems of safeguards" has already acquired rather special connotations. It is therefore felt that it would be better if possible to avoid it in the draft press statement.'

After further correspondence, a final version – with 'systems of safeguards' removed – was watered down a little more to produce a 100-word media statement. It was never issued, because no journalist ever inquired. Despite growing public disquiet since 1956, the media did not notice the signs that major activities were afoot at Maralinga, overlooking the increases in military personnel and much to-ing and fro-ing. Unless an official media release heralded it, the media seemed to show no interest in the events at Maralinga. Of course, this suited the British test authorities, who consciously sought to maintain secrecy.

The media blackout that descended over Maralinga was extremely successful. Given both the level of previous coverage of the British nuclear tests and the rise of anti-nuclear movements throughout the world, the lack of media activity is conspicuous. Vixen B, a test series that ran for three years and involved hundreds of personnel on site, does not appear to have been covered at all. As Lorna Arnold wrote, 'Outside official circles, very few people apparently realised that Maralinga was used for these experimental programmes, and that it continued to be used after Antler'.

Arnold claimed that the British authorities were 'particularly anxious' not to attract any publicity during international negotiations to limit nuclear weapons testing. Vixen B was the major reason for this anxiety, since it produced nuclear fission, albeit in small amounts, and tested an apparatus that came close to many of the characteristics of an actual nuclear warhead. Vixen B was right on the borderline of international law and may have crossed into illegality. The behaviour of the AWRE authorities at the time suggests that they knew Vixen B was in a grey area and political reasons dictated secrecy.

Intergovernmental moves to find a politically acceptable way to slow the race for nuclear arms had begun in 1958. US president Dwight Eisenhower had proposed that test ban negotiations should begin on 31 October that year, pledging a one-year moratorium on weapons testing, and the Soviet Union had agreed. On that date, the Conference for the Discontinuance of Nuclear Weapons Tests had opened in Geneva. A moratorium on the testing of atomic weapons actually stayed in place until September 1961. In 1963, a permanent partial test ban treaty came into effect.

The Geneva agreement was a complication for the AWRE. The UK weapons authorities had no choice but to comply with the agreement, which was binding on the UK. On the other hand, they had an extensive program at Maralinga and plans to expand it. The easiest thing to do was to behave as though the minor trials were not happening. The attitude of the time was summarised by the Australian chronicler of the tests, John Symonds: 'There is no reason to believe that these experiments could be regarded as an evasion of a Treaty, whatever the outcome of the present Geneva discussions. While there is no need to raise the point specifically in Geneva, there is no need to deliberately conceal it, but no public statement is to be volunteered'. Without public statements, there was no media coverage.

Vixen B did not produce mushroom clouds. The major trials sent clouds of minute particles of debris into the stratosphere (more than 10 kilometres above the ground) and spread fallout of short-lived radionuclides over most of the Australian continent, with some isotopes found as far east as the Queensland tropics. The impact of the minor trials was more concentrated, more geographically contained, yet significantly more dangerous close to the firing site. The main dangers were to people in that geographical area, primarily service personnel and scientific staff who were conducting the tests, Indigenous people who traversed the land around Taranaki during or after the trials and later visitors to the site who may have unknowingly picked up radioactive materials or inhaled dust containing plutonium.

The dangers were grave, although there is considerable dispute about their extent. Lorna Arnold took the view that the people exposed to the tests were not seriously affected by radiation, doses of which she said were well within the guidelines laid down by the International Commission on Radiological Protection: 'The people most affected ... were the Aboriginals, because of damage to their way of life rather than directly to their health. They had no rights and their interest in the land was not realized or respected; but this was, and had been, their general situation and was neither new nor peculiar to the weapons trials'.

Vixen was initially proposed as one kind of test, but it evolved into two – Vixen A and Vixen B – a year or so after its first formulation. Vixen A, the original form of the experiments, used mostly beryllium and small quantities of plutonium. It involved studying how radioactive and toxic materials including beryllium, uranium and plutonium might behave in an incendiary or explosive accident and specifically examined how weather conditions influenced the spread of such materials. The tests involved burning the substances in a petrol fire or electric furnace, or dispersing them by high explosive. Thirty-one Vixen A experiments were carried out at the Wewak site, about 15 kilometres to the southeast of Taranaki.

The Vixen A experiments were troublesome for several reasons, not least because the balloons used to hoist a variety of monitoring devices aloft before detonating the bundles of radioactive materials kept slipping their moorings and heading off into the open sky. After one such incident in July 1959, a balloon was found the next day about 10 kilometres away from the test site. Another escaping balloon was not found. The balloon accidents associated with Vixen A caused major disruptions to the test program. In fact, these problems turned out to be a foretaste of more serious balloon incidents in September 1960 connected with Vixen B.

During a storm on the night of Friday 23 September 1960, seven of the eight captive balloons that had been placed for the start of the Vixen B experiments broke free. This was before the first Vixen B experiment, so the balloons were not contaminated. One of the balloons was discovered at Hungerford just over the New South Wales–Queensland border, about 1400 kilometres from Maralinga, and another at Cobar in New South Wales. The recovered Cobar balloon was found to have a faulty mooring system. Test authorities were worried that the footloose balloons would provoke media reports, but while some media did report the escaping balloons, none gave much detail.

After these incidents, the Australian Department of Defence ordered an inquiry, which was carried out by a senior official from the Department of Supply (and included John Moroney on the committee). It recommended that the use of balloons be restricted and proper safety plans be formulated. The inquiry had found that 'a safety plan ... did not exist ... In consequence, the necessary criteria had not been laid down to ensure the development of such balloon safety and mooring systems and handling procedures that would have avoided any escape from the Maralinga range'.

Vixen B proceeded despite the balloon dramas and ongoing problems with the weather. The health and safety controls for these dangerous new experiments included a larger than normal stand-clear zone and dosimetry gauges to detect radiation to be worn by all site personnel. Those in the forward area close to the firing site also wore radiation protection suits, gloves, boots and full-face respirators. The only personnel allowed to be stationed close to the Taranaki site during Vixen B were British service personnel and AWRE representatives; not even the Australian health physics representative Harry Turner was allowed onto the Taranaki site during a firing. Turner's health physics unit was involved from a distance, however, since the health physics requirements for Vixen B were the most demanding of the tests after the major trials.

The exclusion zone in the first Vixen B in 1960 was within

a radius of 40 kilometres from the place of detonation. This was expanded to 43 kilometres in 1961 and ultimately to 56 kilometres by 1963. This compares with safety radii of between 5 and 25 kilometres for the other minor trials conducted at Maralinga at that time. John Symonds noted that the larger than normal safety zones induced speculation and curiosity. Years later, Moroney told his colleague Geoff Williams that 'there was a genuine concern, albeit small, that one of the Vixen Bs could indeed have gone full nuclear'.

The greatest area of radiological contamination from the Vixen B tests was found in, roughly, a 1-kilometre radius from the firing pads. Arrays of sampling instruments were arranged to the north of the Taranaki firing range, to measure atmospheric dispersal and contamination levels for each Vixen B firing, although bad weather often compromised the readings.

The Vixen B assemblies were deliberately made to fit a definition imposed unilaterally by the AWRE (a definition not ratified by the Conference for the Discontinuance of Nuclear Weapons Tests in Geneva) that they 'did not give a nuclear reaction in excess of ten tons of fission TNT equivalent'. At the Royal Commission, WE Jones, the AWRE's co-ordinator of operations for Vixen B, revealed how the test authorities skirted around the new international restrictions on atomic testing. McClellan asked him, 'In scientific terms there was no difference of real significance if you stayed under the ten tons, but the chances are you would not get caught, is that right?' Jones: 'I suppose that is a way of putting it politically'.

Each Vixen B blast blew the feather beds apart, so they could be used only once, after which they were buried in pits dug close to the Taranaki site. British engineer Raymond Carter, at the site during Vixen B, later asserted that the 'quantities of contaminated debris at the Firing Sites had been much greater than originally planned'. In total, Vixen B scattered 22.2 kilograms of plutonium-239 around Taranaki. Although exact figures have never been established, it was later found that rather than 20 kilograms sitting safely in the Taranaki burial pits (there were 19 pits at the time, and two more were dug during the clean-up operations), it was actually spread all around the site, and well beyond, in particles of widely divergent sizes. When Vixen B was underway, a wire fence enclosed the contaminated areas with 'keep out' signs hung at regular intervals, while maps prepared for site staff showed the areas not to be entered. Most of the fences and signs were removed during the Brumby clean-up operation in 1967, in an effort to return the site to its pre-test appearance. In effect, the visible signs of the British tests were removed, but the invisible and more dangerous residue was not.

Plutonium-239 is a dangerous and persistent substance that delivers ionising radiation – the kind that changes the cells of a living body by knocking electrons out of stable atoms – when ingested or inhaled. It must enter the body to do harm – its weak alpha particles can travel only a few centimetres in air. Plutonium is one of the most toxic substances known, and its modern-day uses are strictly controlled, with licences and transparent monitoring protocols required. The small amount of plutonium-239 used in Australia these days is mostly under the auspices of the Australian Nuclear Science and Technology Organisation, which adheres to safety requirements laid down in the *Australian Radiation Protection and Nuclear Safety (ARPANS) Act 1999*, in addition to the ARPANS Regulations 1999. It is used mostly as a measurement standard and to calibrate radiation detection equipment. It can no longer be blown up or spread around an open landscape. Current protocols for the use of plutonium both in Australia and internationally recognise its dangers, including its potential for terrorism. According to scholar Shaun Gregory, 'It has been estimated that one millionth of a gram of plutonium-239 may be sufficient to cause lung cancer if inhaled ... The fissile core of a single weapon would, if perfectly dispersed so that each individual had one millionth of a gram in his or her lungs, be sufficient to threaten cancer in every single member of the human race'.

When plutonium was blown up on the feather beds at Taranaki, the main safety procedures were an exclusion zone during the detonations and some safety equipment, often not properly used. Later many service personnel sought compensation for damage they claimed to have suffered based on several main factors, including the lack of information they received and the inadequate safety measures in place during the tests, particularly Vixen B.

Plutonium poses two main kinds of health hazards: deterministic (also known as non-stochastic) and stochastic. The deterministic hazard involves a threshold level of radiation exposure above which people may be severely harmed or killed, sometimes quickly – for example, by radiation sickness. The stochastic hazard, on the other hand, is based on the probability of developing cancer or genetic damage, which is directly correlated with dose. There is no threshold, meaning in effect that if a given number of people are exposed to any amount of ionising radiation, the probability is that a statistically predictable number of them will suffer cancer or genetic damage. There is no safe dose; any exposure may cause risk of serious illness.

The best known anecdote of deterministic harm caused by British nuclear tests is the 'black mist' experienced by Indigenous people (see chapter 7). For the minor trials, the stochastic effects have caused the greatest concern. These occur over a long time so may not be apparent for many years after exposure. This means the risks are hard to manage and to compensate, particularly since medical science cannot distinguish if a specific cancer is caused by radiation. The kind of contamination present at Maralinga does not pose a hazard when outside the body. Inside the body, the dangers are immense. According to Dr Keith Lokan, then head of the ARL, plutonium particles

> may remain in the lung for a long time where they may expose the lung tissue and give rise to lung cancer. Alternatively …
> because of its long residence time, this material can very slowly dissolve in the lung fluids. Once it goes into solution … it can cross the boundary between the lung and the bloodstream and

then make its way to the bones, and it is acknowledged as a potential source of some forms of bone cancer.

Given the character of plutonium-239, it seems remarkable now that the AWRE authorities did not name it when they briefed the Australians on the Vixen B test series. In the end, this backfired. The Australian authorities began to wonder why they seemed to be cut off from crucial information and eventually changed how minor trial information was provided. As a result, planning for the Vixen experiments did not proceed smoothly. Without doubt, this was more to do with the associated politics and the increasing restiveness of the Australian Government than what now seems to be the sheer foolhardiness of the experiments themselves. When the British were devising the Vixen series, even the notably compliant Australian authorities were starting to have serious doubts, not least because the tests seemed, even to a superficial examination, close to breaching the nuclear weapons moratorium agreed in Geneva. This jousting with international law was becoming a cause of disquiet among the select few who knew about the tests in Australia, as well as much scrambling among the British authorities to make the minor trials fit the new treaty conditions.

The concern about the tests being made public meant that shipping in about 500 personnel to carry out the 1961 Vixen B tests posed a potential threat to security. Noting this, John Symonds commented, 'A disclosure that an operation of the present magnitude was in hand would be difficult to handle'. AWRE officials first hinted to the Australians that they proposed to use the long-lived isotope of plutonium in May 1959. However, the exact description was revealed only after repeated inquiries. The AWRE wanted to augment the existing Vixen series 'by adding a few burning trials to determine the dispersion of plutonium under representative field conditions'. The AWRE endeavoured to slip these tests under the radar, using the cover of Ernest Titterton's insider soundness. Titterton was explicitly criticised by the McClelland Royal

Commission over the minor trials, in part because he had advised the British to say that the fission yield of the 1960 Vixen B tests was zero when he knew it was not: 'The yield was expected to be small, even very small, but not zero', the report said.

Roy Pilgrim, head of safety co-ordination for the AWRE at Aldermaston, issued the highly confidential and later contentious MEP 1960 safety statement at the end of 1959. The 1960 experimental program mentioned Vixen B for the first time, but without specifics: 'Vixen B firings will use long lived radioactive elements including fissile materials. In some rounds the possibility of a fissile reaction is envisaged but the quantity of fission products which would be produced is not radiologically significant compared with the parent material'. A short, handwritten list at the end of the document contained the names of people to whom copies should be sent, including Titterton. The report of the Royal Commission noted that Titterton 'did not advocate a further formal approach through the Commonwealth Relations Office because, inevitably, detailed questions would be asked about the precise nature of the experiments, and how they differ from those already approved'. The sorry saga of how the British tried to dupe the Australians about the true nature of Vixen B had a finale of sorts in London in 1985, as Noah Pearce felt the hostile glare of the Royal Commission. Peter McClellan put it to Pearce, 'You would agree that the 1960 [Vixen B safety statement] document certainly was a totally inappropriate basis on which to form a judgement as to the safety of what was proposed?' Pearce had no choice but to meekly agree.

The tussle over Vixen B between the UK and Australia was only one side of the issue. For the AWRE authorities back at Aldermaston, the main problem was events in Geneva. A 1958 document prepared at Aldermaston by the AWRE's senior superintendent of weapons assembly AR Bryant under the direction of the assistant deputy director of the AWRE Admiral PWB Brooking shed light on this. In a top-secret memorandum titled 'Maralinga Minor Trials in Relation to a Ban on Nuclear Testing', distributed only to three other people including Brooking, Bryant acknowledged that increasing evidence showed most of the minor trials could be safely carried out in the UK, except for

> the precise wording of a statement given by Lord Salisbury to Parliament, which in fact bans firings at Foulness [the AWRE test range on the Thames estuary] using hazardous materials, even in amounts so small that the experiment as a whole involves no hazard. This illustrates the importance of precise phrasing and definition in any policy ban imposed internally in the United Kingdom.

In the UK House of Lords on 7 April 1954, the marquess of Salisbury, the lord president of the council, had stated, 'I can say definitely that no nuclear explosions have been or will be made, nor will experiments be made with fission products or any other hazardous radioactive material'. According to the Royal Commission, this statement was often cited 'as constituting an unbreachable veto on the use in Britain of radioactive materials in explosive nuclear experiments'. Bryant's memorandum suggested two definitions of minor trials to ensure no apparent conflict with the international ban on nuclear testing. The first was a trial in which 'small amounts of radioactive or fissile material are involved in association with the detonation of conventional high explosive in such a manner that no fission results'. The second was a trial in which 'radioactive materials are not dispersed so as to exceed certain agreed tolerance levels outside some agreed radius X miles from an agreed site Y'. Neither of these definitions applied to Vixen B, which clearly produced fission – soon to be acknowledged in Pilgrim's safety statement – and whose radioactive materials were dispersed in plumes far beyond an agreed radius and tolerance levels.

In a top-secret note to Sir William Cook, the AWRE chief scientist leading the British H-bomb project in the Pacific, Brooking explained why he had ordered the memorandum. If Britain had to continue its trials after internationally agreed suspension, as he assumed, it would need 'Ministerial and Australian agreement' on a definition which, 'to be useful to us, must not rule out trials involving RadioActive [*sic*] materials (e.g. Rats) or even fissile material'.

A few weeks later, when Brooking wrote to Cook to put in writing further advice the pair had been discussing, he observed, 'If we are convinced that Maralinga is THE place to do these "unsuspended" trials, then the Australian Government will have to be told or asked', though he noted that this would have to wait until after the Australian elections in November. In the meantime, however, in the light of the international agreement it was essential that the UK and US agree on a definition of what 'we are willing to suspend' so that at the October talks in Geneva both sides would 'talk with one voice'. Brooking also recorded that Cook felt 'radio-active contamination in U.K. is politically impossible'.

The tests continued to take place at Maralinga. Now that the US had removed the decade-long constraints on joint UK–US atomic weapons development and testing imposed by the McMahon Act, they soon took place in Nevada and Los Alamos too. Personal rapport between the UK prime minister Harold Macmillan and President Eisenhower helped the two old nuclear weapons partners to resume their relationship, just as the Soviet launch of the *Sputnik* satellite in 1957 raised geopolitical tensions and made rapid Western technological progress more urgent than ever.

Brooking wrote another letter, on 29 September, discussing in more detail the definitions of minor trials doing the AWRE rounds:

> From the purist's point of view it might be taken to rule out 'single point detonation' trials and maybe certain nuclear trials which could give rise to small amounts of fission. We can however argue that such fission is not the intention of the trial and that if we did produce any it would be an accident, which we are, of course, unable to guard against.

This appears to imply that fission could be produced 'accidentally on purpose', after which culpability could be plausibly denied if they were found out. Brooking noted the US intended to do the same 'if their politicians will let them'. And they did, with the Roller Coaster experiments in Nevada.

Justice McClelland took a jaundiced view of this series of AWRE documents, used as evidence during his Royal Commission. 'The disingenuous tone of this debate ... hardly encourages a belief that the Royal Commission has been told the full story of the minor trials.'

AWTSC secretary John Moroney had pointed to potential problems with plutonium from the Vixen B experiments in November 1963. He wrote to Roy Pilgrim, setting out his concerns about the test range:

> »   Residual plutonium will continue to constitute the
>   predominant radiation hazard at the Range.
> »   There are a few areas, which we believe to be well
>   protected, in which the plutonium levels could constitute a
>   serious radiation hazard.
> »   The present residual plutonium contamination at Maralinga
>   will continue to be a potential hazard for many years and far
>   beyond the period for which activities at Maralinga of the
>   MEP type can be envisaged.

Moroney correctly pointed out that the main hazard at the site was from inhalation of the dust. 'Experience at Maralinga indicates that plutonium moves quickly into the top few millimetres of soil; we do not know how deep it will move ultimately but in a low rainfall region such as Maralinga it may not go far.' He suggested that using the Maralinga site to conduct experiments to assess the hazards of soil-borne plutonium would be a good use of the facilities.

In a memorandum the following year Moroney complained that Pilgrim had never replied to this letter. Eventually, Pearce replied, addressing some of these concerns as plans progressed for the first clean-up operation, known as Hercules, in 1964: 'This clean-up operation has, of course, precipitated the programme on which we were engaged as a result of your letters of November 1963'. Pearce affirmed that the UK had no intention of repatriating any of the Maralinga plutonium, or 'radioactive sources', as he described it, to where it came from, 'and so [we] have the option of disposing of [the radioactive sources] in Australia or of burying them at Maralinga'.

One well-known eyewitness account of the Vixen B trials came from Avon Hudson. Hudson, a member of the RAAF, came to Maralinga in 1960, at the start of the Vixen B test program. Years later he campaigned for recognition of the suffering of the nuclear veterans. He became the first Maralinga veteran to speak to the media about his knowledge of plutonium waste at the test site, gave evidence at the McClelland Royal Commission and also co-wrote a book with Australian academic Roger Cross on the Maralinga legacy, *Beyond Belief*. He told a tale of lax health procedures and pressure to carry out dangerous orders while the experiments were underway. Hudson helped to build the feather beds that held the plutonium-filled assemblies: 'These firing platforms were the ones that could cause so much havoc when it came to spreading radio-active pollution on the range. We knew nothing of what we were doing at the time'.

The minor trials, and particularly Vixen B, were disastrous for Australia. The main substance used, plutonium-239, is among the most toxic materials known, with a radioactive half-life of more than 24 000 years and the capacity to kill people through stochas-tic radiation effects inside the body. The nature of the experiments themselves, where simulated nuclear warheads were detonated on the open range using conventional explosives that blasted radio-active material high into the air, where it spread out in 150-kilometre plumes, was self-evidently dangerous. The experiments were conducted in the presence of hundreds of service personnel, some of whom ventured into the blast area within 20 minutes of an explosion while wearing only basic protective clothing. The experiments were conducted without the kinds of safeguards and monitoring that would enable analysis of risk and causation. The radioactive residue of the experiments was allowed to remain at the site for decades, without robust safeguards and, for a period, without patrols to keep sightseers or Indigenous people away from the contaminated areas.

When Britain finished its testing activities at Maralinga, at the conclusion of the MEP in April 1963, the highly dangerous aftermath of the minor trials lay openly on the ground or just below the surface. The British Government and test authorities knew the damage they had left behind. The Australian Government allowed this to happen through both omission and commission. They created the AWTSC to oversee Australian interests. But for much of its life, it was run by Professor Ernest Titterton, who, in the words of the Royal Commission, 'aided and abetted' British behaviour that was 'characterised by persistent deception and paranoid secrecy'. So just exactly how did the AWTSC look after Australian interests?

6

The Australian
safety committee

*We have not consented to any tests being conducted except under the
strictest conditions of safety, and we do not propose to do so.*

Prime Minister Robert Menzies, 1956.

*The Safety Committee's role was as much concerned with
public relations as it was with scientific safeguards.*

Tim Sherratt, 'A political inconvenience: Australian scientists at the
British atomic weapons tests, 1952–53', 1985.

*It is inconceivable, especially in the light of Titterton's cavalier
treatment of the truth throughout his testimony ... that he did
not know that he had been planted on Menzies.*

Royal Commission into the British Nuclear Tests in Australia,
Report, 1985.

Professor Sir Ernest Titterton's life came to a tragic end. A car
accident not far from his Canberra home in September 1987,
when he was 71 years old, made him a quadriplegic. He lived for
another three years, longing for euthanasia, until an embolism took

his life. The accident came two years after the McClelland Royal Commission report trashed his reputation for his role as the chair of the AWTSC and publicly associated his name with the fictional nuclear maniac Dr Strangelove from the eponymous 1964 film.

The AWTSC is one of the most alarming aspects of the British nuclear tests in Australia. The committee formed to protect Australian interests instead enabled the unfettered ambitions of the British nuclear elite. While it is unlikely that it was set up with this intent, its role evolved to become more of a public relations mechanism than a true overseer of highly dangerous activities. So much of this came down to the personality of Titterton, a compact, trim, crinkle-haired dynamo and divisive figure throughout. Titterton was involved with planning for the tests in Australia almost from the start. When virtually no-one other than Robert Menzies knew that the British planned to test atomic weaponry in Australia, Titterton was edging towards a crucial role – the nexus between the AWRE and everyone else.

Titterton was an English physicist whose career got underway just as the science of nuclear weaponry was progressing dramatically from theory to practice. He joined the wartime British mission to the US and designed the triggering device for the Trinity test at Alamogordo, the epochal test that demonstrated success for the Manhattan Project. He seems to have become a true believer from an early age and carried his convictions of the rightness of atomic weaponry – and of civilian atomic energy – through his life. The AWTSC might have become a well-regarded, conscientious and august body that looked after Australian interests under difficult circumstances had Titterton not commandeered it for more questionable purposes. Now, it looks like a smokescreen.

In an obituary in 1990 the ANU nuclear physicist Trevor Ophel wrote, 'While he gained fame as having "pushed the button" to initiate the first atom bomb test at Alamogordo, the consequences of his time at Los Alamos were more profound. It made him a member of an old boys' network of virtually every leading nuclear physicist, both experimental and theoretical, in the Western world'. He came to the nuclear tests in Australia with existing connections that guaranteed insider status rather than disinterested objectivity.

The AWTSC had two distinct phases, between 1955 and 1957 under chief defence scientist Professor (later Sir) Leslie Martin from Melbourne University and from 1957 to 1973 under Ernest Titterton. Titterton was on the original committee with Martin. The other members were Alan Butement, chief scientist employed by the Department of Supply; Dr Cecil Eddy, director of the Melbourne-based Commonwealth X-Ray and Radium Laboratory; and Professor Phillip Baxter, vice-chancellor of the University of New South Wales and deputy chair of the Australian Atomic Energy Commission. Baxter had played a role in the Manhattan Project, as a chemical engineer based at Oak Ridge, Tennessee, where he had worked on the production of fissile uranium. Martin was a reliable and sensible university and defence man. Titterton was another proposition.

When Titterton became AWTSC chair in 1957, the committee was reduced to three members, narrowing its base to accommodate his larger personality. Alongside Titterton were LJ Dwyer, director of the Commonwealth Bureau of Meteorology, and Donald Stevens, the new director of the Commonwealth X-Ray and Radium Laboratory (Eddy had died the previous year). Titterton remained AWTSC chair until it was reconstituted as the Australian Ionising Radiation Advisory Council (AIRAC) in 1973, at which point he was not invited to continue.

Titterton came to Australia in 1950 after Professor Mark Oliphant invited him to become foundation chair of nuclear physics at the ANU. Oliphant, who had supervised Titterton's research at the University of Birmingham in the 1930s, planned tempting new research in nuclear and particle physics at the ANU. Oliphant was a scientific insider in the wartime race to create a nuclear weapon, but he later became an opponent of such weapons when he saw the human toll from the bombs dropped on Japan. He fell out with

Titterton and was effectively barred from any involvement in the British tests in Australia. Oliphant was actively excluded not just because of his horror at what was unleashed at Hiroshima and Nagasaki, but because he was outspoken about it. The British explicitly requested that the Australians not allow him to be involved. The Americans, who knew him from Manhattan Project days, also did not consider him to be 'sound' – they thought that his views on the open sharing of scientific information made him a security threat. A message to the British High Commission from the British Government, quoted by the Australian journalist Robert Milliken, said in part, 'Oliphant is unquestionably talkative and would give the impression (whether true or not) that he was in possession of all the secrets. It is therefore in the general interest that he should be "kept away"'. In the early 1950s, the Americans declined to issue him a visa to go to a physics conference in Chicago, a moment of humiliation for him. He had no official voice in the conduct of the tests in Australia. All he would do was speak out about his own disquiet. Oliphant would likely have made an excellent safety committee chair, but he was not allowed anywhere near the safety committee.

Oliphant's former pupil did not share his qualms about nuclear weaponry. Titterton, described by ARPANSA scientist Geoff William as a 'creature of the British atomic weapons testing establishment' had, in addition to his Manhattan Project credentials and expertise in high-speed electronic triggering mechanisms, contributed to the US bomb tests at Bikini Atoll in the Pacific. He had then taken up a position with the UK AERE at Harwell. He did not work directly with William Penney, but the pair operated in the same broad circle. Titterton's career record shows he was committed to the development of Western nuclear weaponry. According to his ANU colleague Professor John Newton, Titterton,

> unlike some of his contemporaries, felt no guilt regarding his
> part in the development of these weapons ... He was of the

opinion that it was much better that the Allies first produced them rather than Hitler's Germany, that their use in Japan had saved many US and Japanese lives, and that fear of their use had kept, and would most probably continue to keep, the peace between the major powers.

In the mid-1980s Titterton strongly defended his role at Maralinga when questioned by the Royal Commission and bristled about accusations that he had denied information to the Australian Government, pointing out he had been subject to US and UK secrecy agreements. He continued defending his actions, even as media stories dissected the British tests in the wake of the Royal Commission and found that his actions had been questionable at best (see chapter 10). The Royal Commission report, a document that displayed uncommon levels of ironic humour and controlled outrage, mentioned Titterton's 'special relationship' with the AWRE in several places. It found Titterton's role at Maralinga was to be the AWRE man on the ground, and thereby to limit the information provided to the Australian Government. Justice James McClelland later wrote that 'it would be hard to imagine anyone less suitable than Titterton to be entrusted with a task which called for disinterested concern for the safety of the Australian population from nuclear radiation'.

The report criticised Titterton to such an extent that it led to suspicions that the process was a political witch-hunt. Academic Graeme Turner described the Royal Commission as 'a spectacle of national revenge'. If this was so, the focus for much of the revenge seemed to be on one of its most prominent participants. An account of Titterton's career written two years after his death by John Newton told a somewhat more sympathetic story. In disputing the criticism contained in the commission's report, Newton said, 'The statement that Titterton was "from first to last, 'their man'" rejects any other interpretation of his actions. It appears contrary to the attitude that the Commission adopted in other cases'.

ANU colleagues, many of whom knew Titterton well, consistently defended him, while acknowledging his shortcomings. Trevor Ophel observed that 'rarely has it been more evident that the past is the proper territory of thoughtful historians. Hindsight, conditioned by political and scientific changes evolving over a thirty year period, cannot and should not be used to judge the past'. Ophel noted that Titterton had been 'accused of near treason' by the Royal Commission. It is striking to see how far his reputation deteriorated, from respected scientist and confidante of the British nuclear weapons establishment to Australian pariah.

Nevertheless, a certain relish for the battle can be detected in McClelland's description of Titterton in the witness box 'as a sort of Dr Strangelove figure. So gung-ho about all things nuclear that he gave me the impression that radiation was nothing to worry about and could almost be considered good for people'. Years after Titterton's death, McClelland's assessment of the former head of the safety committee had not softened; he said Titterton was 'totally obsessed with nuclear physics'.

Robert Milliken noted that at the time of the Royal Commission, Titterton and McClelland 'were then aged 69, robust, vain and possessors of particularly sharp and competitive minds'. They were bound to clash, he asserted, because they were so similar.

Whatever the rights and wrongs of the case against Titterton in the Royal Commission, the man certainly made a strong impression on people. Nearly everyone who came into contact with him had a story to tell, not all of them flattering. The overall impression from the many accounts is that Titterton was larger than life, an insensitive and opinionated man with considerable intellectual snobbery who crashed through any barrier. Sometimes he crashed through physically. ARPANSA scientist Peter Burns told of working for the AWTSC in an office next to the front door, 'and you would know when Titterton had arrived because the front door would rocket back on its hinges and smash against the back wall, and he would stomp up the stairs snivelling and snorting. At the end of the day he would stomp down the stairs and smash the door open to make his exit'. He was well known for his pettiness, refusing to install a light in the nuclear physics laboratory 'on the grounds that someone might sit there and read the newspaper', said Newton. Rather than giving away the produce of his home garden, he sold his tomatoes and lettuces to colleagues. This man had once been in charge of the flow of all atomic test information from the British to the Australian Government. As Peter Burns put it, 'He knew all the British weapons people and they told him what they were doing and he told the Australian Government that it was all right. A lot of information closed off there'.

William Penney maintained later that the safety committee was not denied important information. His statement to the Royal Commission said that, in his experience:

> no information pertinent to the possible effects of the
> explosions on the Australian people was withheld. Design
> details of the weapons were not given for two reasons. One was
> simply the expressed policy of the Australian Government not
> to become a party to such information nor to become a nuclear
> weapon state. Secondly, the UK then as now, was obligated
> not to release to a third party certain information entrusted
> to it by the Americans. This did not preclude the Australians
> from receiving military and civil defence information about
> the effects of atomic weapons nor from participating in
> measurements to that effect.

In some ways, the diffidence shown by the Australians – on the one hand actively seeking to be party to the information yielded by the British tests while on the other kowtowing before the British secrecy agenda in the most abject way – did make for a confusing set of messages. Australia did, for a while at least, want to be a nuclear nation, and it wanted to know what the British discovered. The government hoped that if it supported the trials with money and other resources, the British would share this scientific information. The British never thought the same thing, perhaps because the Australians were not upfront about what they wanted. Perhaps, too, the wall created by Titterton was too high for the Australians to climb over.

The AWTSC was formed on 21 July 1955, during the preliminary surveying and commissioning of the Maralinga test range. It arose from discussions between Prime Minister Menzies, Minister for Supply Howard Beale and Minister for Defence Sir Philip McBride. The committee's role was always ambiguous. The British test authorities did not seek its formation. Rather, the committee was set up because the Australian Government wanted to play a distinct part in the tests, to do more than just provide the venue. Ostensibly, the AWTSC was established to evaluate safety aspects of the tests from 1955 onwards, and to act as a conduit between the UK test authorities and the Australian Government. In reality, at least according to some, it did not play a significant role in advising on safety matters. Instead, it provided reassuring window-dressing. The safety committee sought legitimacy for itself in various ways. In October 1958 it published a journal article stating it 'was responsible for ensuring that chosen firing conditions could not lead to damage of life or property on the continent'.

Significantly, the AWTSC did not get involved with the minor trials and concerned itself mostly with the big bomb tests. John Moroney, its former secretary, confirmed this during his feisty evidence to the Royal Commission. He said, 'The minor trials were looked at in the light of the adjective [minor]'. A change occurred later, he asserted, when 'the minor trials ceased to be minor', because of the use of plutonium-239 in Vixen B.

The committee reported to the prime minister, through the minister for Supply. Menzies, in a letter to his minister for Civil Aviation (later Defence minister) Athol Townley in May 1955, specifically requested that the new safety committee 'include members who are sufficiently well-known to command general confidence as guardians of the public interest, and who are not in any way to be identified as having an interest in the defence atomic experiments'.

Right from the start, Australian access to the scientific knowledge bonanza arising from the bomb tests was politely but firmly deflected. Before he formed the AWTSC, Leslie Martin, as defence scientific adviser, was granted some access to Hurricane, on strictly limited terms. The British, while on the surface agreeing to share some information, were rather passive aggressive in denying that information. The Australians tiptoed around gingerly, not daring to suggest access to anything more than peripheral facts. They knew that Titterton was on the inside, as a letter from the chief of naval staff to the Defence secretary Sir Frederick Shedden showed: 'As you know Dr. Titterton is attending apparently by private arrangement'.

Whether Martin and Butement could also be involved was the subject of considerable discussion, and the British High Commission in Canberra stalled on making a decision. A draft Australian Government letter to be sent to the High Commission to try to clarify the situation pushed for the inclusion of Martin, because of

> the considerable Australian contribution in resources to the Monte Bello tests and their close proximity to Australia; the important programme of Defence Research and Development being undertaken by Australia in conjunction with the United Kingdom; and the need for the Defence Scientific Adviser to acquire the fullest information to assist him in advising, from the Australian viewpoint, on the technical feasibility of the use of the Woomera region for future tests; the Australian Defence Scientific Adviser should be present at the Monte Bello tests in order that he might be fully acquainted with the details of the tests.

The Australians reassured the Brits that 'Professor Martin is appropriately covered by the Official Secrets Act, etc ... and he is the Defence Scientific Adviser and comes within the provisions of the Crimes Act which corresponds, in part, to the British Official Secrets Act'. The government had to lobby for one of Australia's most trusted defence scientists to be included in the official party for Hurricane, while Titterton was there 'by private arrangement'.

The tone is conveyed in a series of letters between the Prime Minister's Department and the British High Commission. The secretary AS Brown wrote to Ben Cockram at the High Commission in September 1952:

> My understanding is that the United Kingdom intended by their invitation … that Professor Martin's attendance at the Monte Bello Test would be in such a capacity as would enable him to acquire the fullest information on the details of the test relating to weapon effects and the layout of the site. It was not intended that he should have access to the weapon itself nor its intimate functioning. The Australian authorities agree that at this stage we are not interested in the weapon itself but only in its effects and the general set-up of the test.

Cockram replied:

> I have been instructed to inform you that full details of all weapon effects and the layout of the site will be given to Professor Martin. The accommodation position, however, necessitates that all persons attending the test should be allotted definite tasks, and the health physics team was suggested as being related to the one in which Professor Martin is interested. This suggestion was not intended to limit in any way the undertaking that Professor Martin would be given full access to the information mentioned above. I have also been asked to state that Mr. Butement's attendance is understood by the United Kingdom authorities concerned to be on a similar basis.

The Australians seemed unsure of what to do about the limited nature of Australian involvement. The uncertainty led to paralysis. Martin was stationed uneasily away from the centre of the action. Only Titterton was admitted to the inner circle.

Titterton's involvement was no twist of fate. The British did not wish to leave Australian decision-making to chance. The best way to ensure that Australia was guided to the correct decision was to put a sound man in the middle of the process and make him the obvious choice for the Australian safety committee. The upper echelons of the British establishment felt comfortable with Titterton, who gained access to the tests and, in some cases, the data they produced. This did not extend to others on the Australian side. Martin and Butement did not find favour with the éminence grise of the British tests, Lord Cherwell (chief adviser to Winston Churchill), who actively tried to ensure that they did not receive any significant information. As Milliken stated, 'Cherwell grudgingly cabled his approval of [Butement's attendance] only at the last minute, a fortnight before the Hurricane blast, on strict condition that Butement would have no access to vital efficiency data'. Martin's access to Hurricane was even more disputed and stalled, although in the end he was granted approval largely as a goodwill gesture. Penney wrote to Cherwell, 'We have not treated the Australians very generously in the way of inviting their scientific help. The invitation of Professor Martin would, I think, give them pleasure and would make them feel that we were not attempting to use their land but at the same time keeping them out'.

In 1956, just as the Maralinga site was about to become functional, Titterton published a book on nuclear power and weaponry, *Facing the Atomic Future*, intended for a broad audience. The book named him as professor of nuclear physics at the ANU but did not mention his role on the AWTSC or give any indication that he was involved in the British tests. In the book, Titterton, in views that appear to be at odds with his secretive behaviour, put forwards the need for public information about these issues. 'Insistence on the desirability of informed public opinion on atomic energy matters follows from the basic belief that democracy functions best when the people understand the issues.'

Titterton reinforced his image as a disinterested scientific observer watching the tests from a distance when he participated in some stage-managed media activities ahead of the Mosaic tests at Monte Bello and the Buffalo tests planned for Maralinga. On 15 and 16 May 1956, *The Age* and the *Sydney Morning Herald* simultaneously ran an identical feature in which he answered questions about the test program. In response to the first question, about the purpose of the forthcoming shots, Titterton answered in part, 'If we should ever again have to call on our armed services to defend our freedom it is obviously of the greatest importance that they be equipped with weapons at least equivalent to, and preferably better than, those of a possible adversary. It would indeed be morally wrong to ask them to answer such a call unless we were prepared to so equip them'.

He also addressed rising public concerns about the fate of Japanese fishermen whose vessel had ventured too close to an American hydrogen bomb test in the Pacific, resulting in serious illness and the death of one man. Titterton responded:

> The accident to the fishing vessel *Fukuryu Maru* was most
> unfortunate but it must be remembered that she had strayed well
> into the restricted danger area and also the weapon exploded
> on that occasion was of very large yield – one of the biggest
> ever likely to be tested. The Minister for Supply (Mr. Beale)
> has stated that no hydrogen bomb will be fired in Australia
> and it was recently indicated that the weapons tested would be
> 'small' relative to the American one which led to the accident to
> *Fukuryu Maru*.

The AWTSC met regularly and considered a set range of matters at nearly every meeting. These included air sampling, water sampling, biological samples, meteorological reports and, in the case of the Monte Bello Islands, various kinds of maritime reports. An ASIO operative attended the committee's third meeting held at the University of Melbourne on 28 November 1955, but on the attendance list, the name was later physically cut out of the document. A handwritten note said, 'Name of ASIO official deleted in accordance with S92 of ASIO Act'. This particular meeting dealt with the first ever airdrop of a nuclear device as part of Operation Buffalo at Maralinga in September 1956. The meeting approved this important test 'subject to agreement on location and form of fall out pattern'.

Even at the time, Titterton's apparently seamless elevation to the top AWTSC position caused some uneasiness in the prime minister's office. One member of Menzies' staff wrote in a memorandum, 'To my mind Mr. Beale's proposed committee becomes a one-man band'. Since that one man was known to be a dogmatic pro-nuclear weapons advocate, there was little doubt about the direction of the committee. Nevertheless, Beale signed off on the change. It was a momentous decision.

As long-time AWTSC secretary Moroney recalled, 'While the AWTSC had no responsibility on-site, it was often consulted by the Government for advice on operational matters at Maralinga because it was the only Australian agency informed on the scientific aspects of the trials and their likely impact'. Moroney gave evidence to the Royal Commission that 'Sir Ernest Titterton had greater knowledge than anyone in Australia on the major trials and the minor trials'. Officials from the government departments who dealt with Maralinga came to regret ceding so much control to Titterton and ultimately severely curtailed his power, primarily after the debacle when Titterton concealed the extremely dangerous nature of Vixen B.

The Royal Commission report damned the AWTSC again and again: 'The AWTSC failed to carry out many of its tasks in a proper manner. At times it was deceitful and allowed unsafe firing to occur. It deviated from its charter by assuming responsibilities which properly belonged to the Australian Government'. Beale, however, used the AWTSC as armour in the constant battle to prove the tests were safe. He said in a 1956 media release at the time of Operation Buffalo: 'There is, and always has been, complete unanimity of opinion between the British scientists under Sir William Penney and Australian scientists under Professor Martin as to standards of safety and conditions under which firing should take place'.

The saga of CSIRO scientist Hedley Marston best exemplified the AWTSC attitude to criticism. Marston was one of the few Australian scientists to raise questions about the British nuclear tests while they were underway. This pitted Titterton and Marston, both strong personalities, against each other. Marston was not a particularly pleasant individual, but his name has gone down honourably as the person who took on the AWTSC. The tale was well told in Roger Cross' 2001 book *Fallout: Hedley Marston and the British Bomb Tests in Australia*. In essence, Titterton attempted to stop Marston publishing results that showed alarming levels of radioactive iodine in livestock. Extrapolating (perhaps too enthusiastically) from his findings, Marston asserted that dangerous radioactive fallout from both the Mosaic tests at Monte Bello and the Buffalo tests at Maralinga was blowing through countryside, towns and cities, entering the food chain via animals and from there reaching millions of unsuspecting people.

Marston was an agricultural scientist whose job was to determine the amount of radioactive iodine in the thyroids of sheep and cattle in the fallout area of the 1956 test programs. In fact, he took samples from way outside the originally agreed test area and showed that fallout was entering the bodies of animals in most parts of Australia, as far away as Townsville on the north Queensland coast. Marston did not test specifically for the more dangerous strontium-90, because in his view (contradicted in more recent years) iodine-131 was a marker for it. Marston argued that the national school milk program then operating in Australia seemingly guaranteed uptake of this deadly radioactive isotope, leaving children with lifelong radioactive material in their bones where it could potentially cause bone marrow cancers. Titterton disagreed vehemently with this interpretation. After Mosaic G2, the AWTSC had reassured Menzies that all contamination measured on the mainland was low and nothing to be concerned about. Marston's measurements did not support this view.

The AWTSC tried to shut Marston down, placing pressure on the *Australian Journal of Biological Sciences* not to publish his article. Dr (later Sir) Frederick White, then chair of CSIRO, Marston's employer, attended its 19th meeting on 11 June 1957, where an agreement was struck on censoring Marston's article. This also happened to be the first meeting of the newly constituted AWTSC under Titterton's leadership. White, effectively the meat in the sandwich, was provided with a string of demands from the AWTSC that included its clearing the final version of the paper and removing 'personal attacks and unsubstantiated opinions'. The committee also insisted that several graphs containing data on gamma ray fallout be removed altogether, on Penney's direct orders, although Marston was not to be told that Penney had insisted on this. When an abridged version of the article was published in August 1958, it was scarcely noticed, and Marston faded into the background.

The two biggest failings of the AWTSC were the huge but undisclosed yield of Mosaic G2 at Monte Bello in 1956 and the devastating plutonium contamination caused by Vixen B. At Hurricane and Totem, the Australian presence had been less than peripheral, and the lax safety at both series had the potential to cause significant political problems. The Australian Government wanted a proper mechanism to ensure a role in the subsequent tests. On 20 June 1955 Menzies agreed in principle to the British going back to Monte Bello for Mosaic, on the understanding that safety of people and animals in the vicinity of the tests would be ensured. The AWTSC was set up within a month, and the committee had precious little time to prepare for the biggest British test in Australia.

By now, test insiders knew the inadequacies of the safety preparations for Hurricane and Totem. Some of the many safety problems included the inability to predict fallout when the winds were strong, the effect of meteorological conditions generally on radioactive contamination and estimating the height of the mushroom cloud. The British meteorological expert Albert Matthewman prepared a report outlining these uncertainties. In evidence to the Royal Commission, Matthewman acknowledged that in many cases predictions did not match theoretical modelling. Much of the blame for this, he said, was the extreme difficulty in forecasting the wind: 'Nine tenths of the explanation of the departures [from the modelling] were due to errors in the forecast of low level winds'. Clearly the safety foundations for nuclear testing in Australia were not strong, and it is hard to find evidence that safety was ever a top priority for the AWRE. Catching up with America and the USSR took precedence.

At its first meeting in July 1955, the AWTSC put on a show of strength. The new members agreed that neither the Mosaic tests scheduled for mid-1956 nor the Buffalo tests at the new permanent range due to take place in September 1956 should be given the go-ahead without proper information to better calculate the hazards. The committee held urgent meetings with AWRE visitors that same month, in an attempt to get on top of problems already evident in Mosaic and Buffalo. Monte Bello was known to be meteorologically difficult. In fact, the Hurricane test had been restricted to October, the only time when the weather was reasonably predictable. Yet Mosaic was to take place in May and June because the British were in a hurry to test the triggering mechanism for their new hydrogen bomb. At this time the prevailing westerly winds at Monte Bello were almost guaranteed to send the radioactive cloud across the mainland. But when the Churchill government decided to develop the hydrogen bomb in 1954, following the lead of both the US and the USSR, there was no time to waste. The British imperative to schedule the Mosaic series threw safety planning into panic mode.

Reports were prepared and experts consulted. By September 1955, the AWTSC knew that the meteorological conditions for the planned Mosaic firings were not likely to be safe. The problems intensified when a new report indicated that the seas were likely to be rough too. Members of the Seaman's Union, which opposed the tests, were suspicious when the AWTSC asked its members to wear film badges to detect radiation. (Unfortunately film badges were rarely analysed. They were often worn and then discarded, without proper investigation of what they showed.) On 1 June Martin wrote to Beale to explain the rationale for this precaution:

> The Safety Committee does not expect any dangerous fallout on ships and will in fact take all precautions necessary to ensure that there is none. The film badges are insisted on only because they will provide the best evidence to refute any allegation of the occurrence of high intensity fallout, that might be made either maliciously or from the misinterpretation, improper use or faulty operation of monitoring devices.

At its third meeting in November 1955, the AWTSC discussed the safe firing of the Mosaic tests, though it still did not have all relevant information. AWTSC members, all due at Monte Bello for the tests, left for the long trip to the northwest intending to meet there once the material was supplied. While these meetings are presumed to have taken place onboard the HMS *Narvik*, one of the British ships on site, no official minutes were kept.

A complication arose in the lead-up to Mosaic because the AWRE decided unilaterally to define radioactive contamination using the neutral nomenclature of A and B, rather than the previous descriptive designations of zero risk and slight risk. Charles Adams, the AWRE's chief of research, said in a memorandum informing the AWTSC about the change, that 'the new definitions ... have received very careful thought here and we advocate the different nomenclature since we believe the term "slight risk" implies a greater possibility of damaging effect than is justified by the definition'. The AWRE wanted the Australian committee to agree to the use of a contamination level halfway between A and B. Adams continued, 'The adoption of half level B as the criterion would, in our opinion, give only the slightest chance of any physiological effect, which in any case would be temporary and only just observable. We should, of course, hope for forecasts which will show that exposures will be below this level'.

This change left room for speculation about exactly what level of contamination was acceptable, and whether the same levels were also acceptable for Indigenous people living a traditional lifestyle, whose circumstances were different. For a start, in many cases they were barefoot and semi-clothed or naked. Half of level B accorded with the standards adopted by the US at its Nevada tests. However, the AWTSC objected that the US standard was not necessarily suitable for Australia and said that half level B would be accepted only for the Mosaic tests, not for those at Maralinga, where the concentrations could be expected to build up with each successive test. In the end, and after a struggle, the highest acceptable dosage for Aborigines and others was set at level A.

Leslie Martin believed that the safety committee should always be present for bomb tests. The attendance of all members at Mosaic G1 was agreed at the fourth meeting of the AWTSC on 10 January 1956:

> The Chairman was of the opinion that the whole of the
> Committee should be present at the trials and as it was not
> feasible to accommodate them on *Narvik* for any length of
> time, they should be located at Onslow with the rest of the
> mainland party. The Secretary was instructed to arrange for the
> accommodation of up to seven Safety Committee members at
> Onslow and to provide 2 vehicles for their use. Fast transport was
> also required to the Islands when the tests were on.

Members of the committee arrived at Onslow on 14 May, a one-pub town on the mainland that was the departure point for test personnel to get to the Monte Bello Islands. They were quickly transferred to *Narvik* to witness the test and hurriedly met to discuss some of its worrying aspects, notably the enigmatic but definite hints about 'light elements' being used as 'a boost'. This seemed to suggest a fusion weapon, something that the UK prime minister Anthony Eden had mentioned in a cable to Menzies the year before.

Light elements, particularly lithium and the isotopes of hydrogen called deuterium and tritium, are associated with fusion (thermonuclear) bombs, while the heavy elements uranium and plutonium are associated with fission bombs (although heavy water containing deuterium also was used for early fission weapons, for different reasons). However, it is possible to increase the explosive power of a fission bomb by initiating nuclear reactions among light elements to raise the temperature at the centre of the bomb quickly and by a great deal. The Royal Commission noted that:

> the Committee members wanted to know which elements
> would be used and the quantity of each. The information was
> ultimately given by [operational commander Commodore
> Hugh] Martell and was said to have been restricted to their own
> use. What information was given is not recorded but as the firing
> proceeded, they were evidently satisfied with that information.

Mosaic G1, the smaller of the two Mosaic bombs, was detonated on 16 May. The members of the safety committee watched the explosion from the deck of the *Narvik*. Soon after G1, they departed the remote northwest but would be back just over a month later. Mosaic G2, the biggest device ever tested in Australia, would be detonated next.

Early signs of discord between the AWRE and the Australians were evident in the lead-up to Mosaic G2. The climatic conditions were so fragile and the window of opportunity so limited that when the AWTSC started to question aspects of the safety of Mosaic, Commander Martell (standing in for Penney, who did not attend Mosaic) was greatly displeased. The Australians believed that the safety committee gave them more power than was understood by the British, specifically the power of veto up to the moment of firing. As John Symonds observed, 'The Operational Commander [Martell] had been given full responsibility for the operation on the UK side. The fact that the AWTSC had power to veto the operation was subsequently a surprise to the Operational Commander and caused him concern during the final stages of the actual test period'.

The AWTSC had been set up to hold such a veto, although exactly how it could actually stop the British carrying out a test was never explained or tested. The Royal Commission later found that despite the effective power of veto, the AWTSC 'was not provided with sufficient information to discharge its function properly for the Mosaic series'. Another cause for both concern and fury for Martell was the firm direction from Howard Beale that the G2 test should not take place on a Sunday, for religious reasons, which was an Australian Government, not an AWTSC, requirement. The AWTSC was under considerable pressure from the Australian Government to ensure that Australia asserted itself in this first test series.

The deadlock was overcome when Martell and Adams provided the committee with a bit more information. G2 was detonated just after 10 am on 19 June. Earlier assurances from the British authorities had suggested that neither Mosaic device would be more than two and a half times Hurricane, suggesting an upper limit of about 62.5 kilotonnes. G2 may in fact have been a 98-kilotonne monster, although none of the Australians, including Leslie Martin, knew this at the time. (It should be noted that 98 kilotonnes remains a disputed figure. While it has been accepted by many authors, such as Robert Standish Norris and Joan Smith, others, such as Zeb Leonard, have suggested that there is no evidence that it was anything other than 60 kilotonnes, the figure recorded by the Royal Commission.) Nevertheless, the AWTSC knew it was big.

Martin prepared an upbeat if somewhat confused report onboard the *Narvik* which was sent to the prime minister. He suggested that the radiation cloud had moved out to sea, and also that some low-level radiation had spread eastwards across the Western Australian coast, towards the centre of the continent. 'From analysis of the detailed data available to us the Safety Committee has satisfaction in reporting that the safety measures were completely adequate. There was absolutely <u>no hazard</u> to persons or damage to livestock or other property.' The report made the rather strange statement that 'the fallout on the NW coast was harmless in the extreme'. The Royal Commission later described this report as 'misleading', and Martin's safety assurances as 'grossly misleading and irresponsible'.

As previously discussed, Beale sent out a frantic 'please explain' to the safety committee at Monte Bello after a miner called Stewart Stubbs at Marble Bar, hundreds of kilometres away, detected high levels of radiation on two Geiger counters on 20 June. He used Geiger counters because he was involved in uranium exploration. In drizzling rain, Stubbs detected radiation that was 'off the scale', according to his testimony at the Royal Commission. He didn't keep it to himself but radioed the airport at Port Hedland and spoke to someone whom he believed to be 'a British scientist'. (He told the Royal Commission that he thought the name of the person was Penney, although Sir William was not at Monte Bello for Mosaic.) To whom exactly he spoke is unclear, but somehow his Geiger counter readings were provided to a journalist, who wrote a story that sparked public disquiet. According to the Royal Commission report, there was no official fallout monitoring station at Marble Bar, and therefore the degree of contamination there will never be known, 'but if drizzling rain occurred, fallout contamination would have been greater than for dry conditions'.

One of the main roles of the AWTSC was to establish a network of monitoring stations around Australia, and eventually 60 stations were created, including in all state and territory capitals. However, their efficacy is open to debate. The safety committee oversaw an almost comical array of sticky paper that didn't work when it rained, air pumps that measured airborne contamination but were regularly clogged by dust, battery-powered dosimeters with batteries that were usually dead, and other totally ineffectual methods for detecting and measuring radioactivity. Even for the times, these methods were inadequate. The 1984 Kerr Report on the aftermath of the British tests found that the 'wearing of film badges (dosimeters) was so erratic and, in some cases, the measuring of doses so arbitrary, that … little weight can be placed on the validity of records as an index of long-term dose commitment'. The report noted, too, that when Mosaic G2 was about to be detonated, the AWRE supplied the AWTSC with 50 sets of fallout deposition monitoring equipment, but they deployed only 28 sets. Whether this was due to lack of time or lack of interest was not noted. The AWTSC also used aerial surveys by low-flying planes to obtain fallout measurements. The Royal Commission found that these readings would have to be multiplied by 10 to accurately reflect the contamination on the ground. But, as author Joan Smith wrote, 'the Atomic Weapons Tests Safety Committee used the actual readings to give assurances to the public about the level of fallout from the tests'. Exactly why is unclear. It's likely they did not want to upset the British or fuel public fears. They were also not in possession of all the information that the British had.

Once the stresses of the highly problematic Mosaic series were out of the way, the safety committee turned its attention to Maralinga, which was just about to become operational. Kittens tests were already underway there, and these notionally fell under the remit of the AWTSC, although, as previously noted, the committee was not really interested in the minor trials. Again, meteorological conditions were discussed at the committee meetings, especially the perennial problem of winds. In the 1950s, there were no satellites to assist in predicting weather. If meteorology is an inexact science now, it was far more so then. Would the swirling air currents send fallout to populated areas, with incalculable risks both physical and political? The AWTSC came up with guidelines to avoid this, largely by limiting the yield of bombs tested at Maralinga to no more than 10 kilotonnes and ensuring that the most precise meteorological calculations possible were made. The British test authorities did not heed any of these recommendations. The committee was also concerned about the presence of Aboriginal people in the area, and the fact that their lifestyle made it more likely that they would be affected by fallout and contamination. At the Royal Commission, William Penney claimed he had been unaware of such concerns.

When Operation Antler, the final major bomb trial at Maralinga, was in its planning stage, there was a fundamental change in the AWTSC. The committee was reduced to three, and Titterton became its chair. The role changed too. The committee was now required to report to the prime minister only on the safety of the weapons tests. A new body, the National Radiation Advisory Committee, took on the task of examining radiation safety in the community. That committee had a broad responsibility to advise the government on all matters concerning ionising radiation in the community, including fallout from the atomic weapons test programs of other countries and not just those related to the British nuclear tests. The chair was the distinguished Australian scientist Sir Macfarlane Burnet, who was joined by Leslie Martin and Alan Butement from the old AWTSC. This new committee had been created on Martin's advice. As defence scientific adviser, he was acutely aware of a backlash that could not be ignored caused by the swift rise in the world's nuclear weapons arsenal and the testing by three nations. Atmospheric nuclear tests sent radioactive elements high into the upper atmosphere and stratosphere, with who knew what long-term effect. The advisory committee was intended to provide advice on how the Australian community might be affected.

The ascension of Titterton to the head of the safety committee was a welcome development for the British, particularly during the trickier than usual negotiations for Operation Antler. The AWRE wished to deal only with Titterton, who held talks with senior

AWRE staff when he visited the UK in March 1957. According to the Royal Commission, he was given details about Antler during this visit that he did not share with the Australian Government.

A typical example of the way that Titterton operated can be seen in the case of cobalt-60. As the 1950s wore on and atomic weaponry started to become a source of considerable public anxiety, rumours often gained legitimacy. The idea of a cobalt bomb – created by adding cobalt to a nuclear weapon, a process known as salting – began in 1950 when the physicist Leo Szilard postulated it as a hypothetical 'doomsday device' that would wipe out all human life. The addition of cobalt would increase the amount of radioactive fallout produced by the weapon, making it more deadly. It didn't actually exist, but the prospect caused some disquiet. Titterton addressed this issue in one of the articles he published in the Australian press during the test series. There was no cobalt bomb on the British drawing board, but, all the same, cobalt later featured in the severe criticism of Titterton during the Royal Commission.

The first Antler bomb at Maralinga in 1957 contained a radioactive cobalt isotope, cobalt-60. Titterton knew this extremely dangerous substance was going to be used but did not tell the Australian officials at the site. Doug Rickard, an 18-year-old technician who was monitoring fallout for the Commonwealth X-Ray and Radium Laboratory, found small pellets of the highly radioactive substance on his routine patrol at the Tadje site after the first Antler test. He scooped some of the pellets into a tobacco tin. The tobacco tin was so radioactive that the radiation readings were off the scale. The Australian health physics representative Harry Turner reported this to the British authorities, who acted to lock down the information. During the Royal Commission, Titterton claimed that he had deliberately concealed information about the use of cobalt as a 'test' for Turner and his health physics team. Jim McClelland's scorn comes through in the transcript. He accused Titterton outright of being a liar.

The cobalt was used during the Antler Tadje test to detect the amount of energy released at the time of the explosion, although it failed in this purpose. In fact, the substance was simply sprayed out from ground zero and fell as pellets all around the test site. Doug Rickard was not meant to discover it. Titterton was, of course, party to the information, but no-one else on the Australian side was supposed to be. According to a top-secret memorandum from Charles Adams to Admiral Brooking back at Aldermaston, 'With some difficulty I obtained permission from the Director to inform Titterton (as Chairman of the Safety Committee) that we intended to use such an indicator. Titterton was entirely sympathetic, raised no difficulties, realised that we were not adding any real hazard, and agreed that the information should go no further on the Australian side'. By informing Titterton, the British could claim that they had done their duty to the Australians, as they did at the Royal Commission, while knowing that he would not pass anything on.

The AWRE brass were concerned about Turner spreading the cobalt information, and a secret memorandum recorded, 'It appears that Turner has reported to [range commander] Dick Durance in addition to reporting here. In doing so I think Turner has misconstrued his terms of reference which were to report in the first instance to A.W.R.E.' Adams indicated that he was urgently trying to contact Titterton. 'If it is necessary to correspond with the Australian Department of Supply or the Range Commander, I should much prefer to do so in terms agreed with Titterton, rather than write indepdently [sic].' In the end, a huge number of cobalt-60 fragments were encased in lead and buried at the airfield, the place where many problems were buried out of sight.

The undoing of Titterton was Vixen B, the most secret and most dangerous of all the Maralinga tests (described in chapter 5). The AWRE told the Australian authorities next to nothing about the minor trials, including the new style of test represented by Vixen B. But, especially since plutonium-239 would be used, the British knew they would have to tell the Australian Government *something* about them. The Vixen B experiments sparked a series of correspondence between the two governments and the test authorities that continued for a couple of years. Some of these letters contained a heated or exasperated tone, as relations between the parties were increasingly strained.

The correspondence began with a carefully worded letter sent by Titterton to Allen Fairhall (who in 1959 was a member of the Parliamentary Standing Committee of Public Works and in 1961 became minister for Supply). Titterton let Fairhall know that the longest lived isotope of plutonium was on the agenda and that this might have some political overtones given the then-current moratorium on atmospheric nuclear tests. The letter, sent on 10 July 1959, was followed on 30 July 1959 by a formal request for approval of the relevant tests, the Vixen A and B minor trials at Maralinga, from the UK minister for Supply to his Australian counterpart. It said in part, 'Although these experiments are in no sense nuclear tests, it will be desirable to avoid publicity for them in order to remove the risk of their being misrepresented by ignorant or ill-intentioned persons'.

The Australian minister for Defence Athol Townley, who had taken over from Philip McBride in 1958, also received a copy of this letter, and the following day he sent a reply notable for its glimpse into the preoccupations of that era:

> I am not troubled very much by the trials themselves ... The
> political aspects, however, can be potentially dangerous ... for the
> first time it is proposed to use explosives on the Woomera Range
> which will bring the usual howl from the 'Ban the H Bomb'
> section of the community – Communist and otherwise. It is my
> view, therefore, that there should be some political discussion on
> it ... I would hesitate to put it into full Cabinet, purely on the
> 'need to know' basis.

This letter attracted the scorn of Justice James McClelland in his 1985 report:

The decision to allow fissile material with a half-life of 24 000 years to be spread on Australian soil, no matter how remote, was evidently in the hands of politicians, one of whom [Townley] did not know that the Woomera Range and the Maralinga Range were not the same thing, and with the exclusion from such a decision of all but two or three members of the Cabinet. This is an instructive little lesson in the style of democratic government in Australia during the Menzies era.

As we have seen, Roy Pilgrim, head of safety co-ordination for the AWRE at Aldermaston, was responsible for issuing the highly confidential and later contentious MEP 1960 safety statement at the end of 1959, the first to include Vixen B. It acknowledged the use of 'long lived radioactive elements including fissile materials', while downplaying the consequences. Pearce admitted at the Royal Commission that the safety statements weren't an appropriate basis for the Australians to judge the safety of the proposed tests. Significantly, the official document requesting Australian approval for the 1960 MEP did not mention Vixen B explicitly, even though it had been discussed with Titterton in his role as head of the AWTSC. Titterton had conveyed the view to the AWRE that 'the approval process already granted by the Australian Government for the series of experiments now at Maralinga in 1960 covered the type of experiment we now wish to carry out' and did not advocate a formal approach, which would inevitably have led to detailed questions.

That was not the end of the matter. The Australian authorities were not happy about being denied specific information on the Vixen A and B tests and began exercising an unprecedented capacity to stall and hinder the test program (albeit temporarily), to the chagrin of the British test authorities. The Royal Commission noted, 'The 1960 proposal for assessment tests, which included the Vixen B tests, caused Australian officials, particularly in the Department of Defence, to question the existing procedures for approval of the program. It was apparent that decisions which demanded political input were being taken by the AWTSC, through its Chairman, without reference to the appropriate Ministers'. The AWTSC was not a free agent; it had been established to represent the interests of the Australian Government and citizens. People were starting to notice that this was not actually how it was operating.

Titterton had gone too far. His attempts to push the Vixen series through the approval stage gave credence to allegations that his first loyalties were to the AWRE and not to his employer, the Australian Government. Australian officials began to have doubts about a process that saw safety information from Aldermaston being sent only to Titterton, whereupon it often stopped altogether. The new and more dangerous kinds of tests on the range made this increasingly unacceptable.

The dynamics at Maralinga were changing, and Titterton's unquestioned status started to crumble. At this point Menzies got involved in the exchange of messages over the political implications of the plutonium tests. According to the Royal Commission:

> When told of the UK proposal, the Australian Prime Minister consulted with senior Departmental officials whose advice contained the warning that Australia had very little information concerning these particular tests. It was not clear to them that the AWTSC [was] any better informed though it was possible that the Chairman had been given some information by AWRE officials.

In reality, Titterton was far better informed than any other Australian official, but, in a conclusion of the Royal Commission, he 'played a political as well as a safety role in the testing program, especially in the minor trials. He was prepared to conceal information from the Australian Government and his fellow Committee members if he believed to do so would suit the interests of the United Kingdom Government and the testing program'.

A letter from the British deputy high commissioner Neil

Pritchard to Maurice Timbs in the Prime Minister's Department attempted to play down Australian annoyance over the Vixen series and smooth away the growing disquiet. 'As I understand that your Government would like some further information about the additional trials, which for convenience have been given the codename of "Vixen B", we have now been asked to advise you as follows.' The letter then set out some basic information about Vixen B in five dot points, including the claim that the AWRE safety statement given to Titterton for the safety committee gave 'details of the likely contamination and of the precautions to be taken'. The letter concluded with a plea for rapid resolution of the remaining issues so Vixen B could be added to the MEP and pointed out that 'the experiments have been agreed by the [Maralinga] Board of Management subject to this formal approval and all precautions are in hand. We expect the United Kingdom Servicemen to arrive early in July'.

The UK Government clearly assumed that the experiments were approved as part of the broad agreements already in place. The British were keen just to get on with it and appear to have been surprised that the Australians were asking questions. In fact, there was no formal approval at this stage. After Allen Fairhall received and reviewed this letter, he wrote to Defence Minister Townley, indicating his support for Vixen B but handing the matter to Townley for the final decision. 'Defence officials investigated the new situation and noted that, outside the AWTSC, knowledge of the trials was limited to the very general comments about them in the UK High Commissioner's note.'

The Australian Government did not approve Vixen B until 18 August 1960 and conveyed this to the UK high commissioner on 30 August. The Defence Department in particular was concerned about how little information was getting through, though it was concerned less about the safety arrangements than 'the possibility of knowledge of the arrangements falling into wrong hands. It was a matter for political judgement how serious any embarrassment stemming from such knowledge might be'. When the Defence

Department finally approved the tests, it included some new conditions for approval of future Maralinga tests: 'The way was then clear for further discussions about some more formal channels of communication between Australia and the UK authorities in addition to the original AWTSC/AWRE channel'.

In the end, Titterton's decision to keep the details of the series secret from his employers sidelined him. The departments of Supply and Defence eventually bypassed him altogether and went direct to the AWRE. The depths to which he had sunk in the estimation of the Defence Department and the Prime Minister's Department was evident in a rather brief and uninformative letter written by the secretary of the Department of Supply John Knott. Titterton had written a lengthy letter to Knott on 24 August 1960 in which he set out the growing uneasiness he was hearing from both the Defence Department and the Prime Minister's Department over Vixen B. Titterton was concerned about a view, particularly of Maurice Timbs from the Prime Minister's Department, that the safety committee did not have enough information on the exact nature of Vixen B to properly assess its safety. 'The [AWTSC] takes a most serious view of this: it reflects on our integrity and suggests that we agreed to trials without knowing whether they were safe or not.' There was a veiled threat in Titterton's letter: 'We would feel it most improper for us to continue in our work unless we can be assured that we have the complete confidence of the Prime Minister's Department, the Department of Defence and the Department of Supply'.

The letter prompted a lukewarm reply from Knott two days later, which said in part, 'May I say at once ... that you and your Committee have the full confidence of the Department of Supply and equally I feel sure would this be so [sic] in respect of all other Departments and officials concerned'. It appears that Supply was the only Australian department that came forwards with the requested vote of confidence. From the other two, there was silence.

The AWTSC was a belated and ineffectual attempt to give

Australia some say in the ground-shaking events taking place on its own turf. The AWRE really became comfortable with the committee only when Titterton took over the chair in 1957. Before that, Leslie Martin had played as straight a bat as he could but scored virtually no runs. Under Titterton, the committee became a sham, giving the appearance of playing a serious role but in fact undermined from within by someone who had no intention of keeping the Australian Government properly informed. But if Australians as a whole were badly served by the safety committee, for one group in particular the British nuclear tests were profoundly devastating.

7

Indigenous people
and the bomb tests

*Long time ago, before whitefellas came, Anangu lived on their lands
for thousands and thousands of years. The land was their life. They
loved the land. They cared for the country. They knew all its secrets
and they taught those secrets to their children and their children's
children, tjamu to tjamu (grandfather to grandson), kapali to kapali
(grandmother to granddaughter).*

Yalata and Oak Valley Communities with Christobel Mattingley,
Maralinga: The Anangu Story, 2009.

*Whitefella sent us away. Whitefella came
to this place and sent us away.*

Tommy Queama from Ooldea, speaking through an interpreter
at the Royal Commission, 1985.

In June 1956 the Department of Supply chief scientist Alan
Butement wrote a furious letter to the controller of the Weapons Research Establishment, an Australian Government organisation based at Salisbury, near Adelaide, that ran the rocket range at
Woomera. Department of Supply native patrol officer Walter Mac-
Dougall had threatened to publicly disclose the potential harm
to local Indigenous people after a weather station associated with

the Maralinga test range was proposed at Giles. The controller, HJ Brown, set out MacDougall's concerns in a memorandum. Butement responded, 'Your memorandum discloses a lamentable lack of balance in Mr. McDougall's outlook, in that he is apparently placing the affairs of natives above those of the British Commonwealth of Nations'. Justice James McClellend described this as 'one of the most telling of all statements to come before the Royal Commission'. Maybe it was also telling that the chief scientist did not know how to spell MacDougall's name.

McClelland found strongly against the British and Australian governments for their treatment of Aboriginal people, not just at Maralinga but also on the mainland adjacent to the Monte Bello Islands and at Emu Field. 'If Aborigines were not injured or killed as a result of the explosions, this is a matter of luck rather than adequate organisation, management and resources allocated to ensuring safety.' The damage done to the Indigenous populations was perhaps the most contested and tragic of all issues relating to the British nuclear tests in Australia. It was colonialism in microcosm and speeded up.

The murky status of Aboriginal people in Australia at the time was at least partly to blame. Aboriginal people did not have the federal vote, they were not counted in the census, and their affairs were not discussed in any depth in the Australian mainstream. They started to get the vote, state by state, from 1962. This makes somewhat fanciful Ernest Titterton's comment to the Royal Commission that if the Aborigines in the area hadn't liked the tests they could have voted the government out. Not until a referendum in 1967 were they included in the census. Shockingly, only a couple of generations ago, in some parts of Australia, Aborigines were administered under state flora and fauna Acts. The 1967 referendum also changed the Australian Constitution to enable the federal parliament to enact legislation on Aboriginal affairs. Before then, legislation concerning Aboriginal life was the domain of state governments.

The treatment of those affected by the atomic tests was not

unusual. It was perhaps more unusual that a government employee, Walter MacDougall, was assigned to look after their interests, given the prevailing view at the time, namely that the kindest thing was to let them die out. They were practically invisible. Invisible people can be harmed with few consequences.

This attitude went right back to Operation Hurricane in 1952. In fact, it was doubtful that more than cursory consideration was given to Aboriginal safety when planning that hastily organised first British A-bomb test. The Royal Commission found that the authorities relied almost entirely on a document titled 'Some Notes on North-West Australia' prepared by the HER scientist ER Woodcock, whose main source was the *Encyclopaedia Britannica*. According to the Royal Commission report, Woodcock recorded a population of 715 people living within 240 kilometres of the Monte Bello test site, 'excluding full-blooded Aboriginals, for whom no statistics are available'. Considerably more information was provided about the hens, ducks, cattle, horses and sheep in the vicinity. There was a paucity of information because Aboriginal people were not counted in the national census. Yet the Royal Commission found that the state government of Western Australia had better information at the time. This information specified that some 4538 Aboriginal people lived in regions on the mainland close to Monte Bello, hundreds in small townships, and the rest spread out at various missions, on cattle stations or in traditional family groups.

Titterton told the Royal Commission that he had been charged specifically with ensuring the welfare of the Aboriginal people near Monte Bello. He testified that Menzies had asked him to 'stick your oar in to make as certain as it is humanly possible ... that there will be no adverse effects on the Australian people, flora and fauna, and in particular the Aborigines. From the first five minutes I was involved a major concern of the Australian Prime Minister was Aborigines'. Yet Titterton did not apparently stick his oar in very deep, and his evidence on exactly how he did this was vague. He seemed to be under the impression that the federal government held such records, when it was only state governments who kept data on Aboriginal people during this period.

Perhaps his overall approach was summed up in this comment: 'The overriding condition was that there would be no significant fallout on the continent. Now, if there was no significant fallout on the continent that can do anyone any damage, you do not have to differentiate between Aboriginals and Europeans'. As the Royal Commission concluded, this failed to 'consider the distinctive life-styles of Aboriginal people'. And since 'no record was made of any contamination of the mainland it is impossible to determine whether Aborigines were exposed to any significant short or long-term hazards'. Aborigines leading traditional lifestyles have far greater connection to the land upon which they walk and the water, plants and animals in their vicinity. Aborigines and Europeans did indeed need to be differentiated.

When the Mosaic tests were held at Monte Bello in 1956, again the welfare of Aboriginal people on the adjacent mainland did not figure in planning. Commodore Hugh Martell, director of Mosaic, dismissed the idea: 'Aborigines are nothing to do with Mosaic ... The question does not arise. There are no Aborigines on the Monte Bello Islands'. But Aboriginal people on the Pilbara coast and inland would certainly have been affected. At no time during Mosaic were they properly taken into account.

For the 1953 tests at Emu Field, there could be no denying that Aboriginal people were in the vicinity. They were there for all to see, if they cared to look. Emu and, later, Maralinga were chosen as locations despite what would these days be considered challenging, if not insurmountable, barriers. Notions such as land rights, informed consent or occupational health simply did not arise for local Aborigines. The land was not uninhabited, but *terra nullius* ('nobody's land', deeming there were no property rights in the continent when the British took it) was accepted in law. The lands were not settled in the Western sense of the term. Rather, people traversed them for hunting, water gathering and ceremonial purposes, as they had for tens of thousands of years. Before the major mainland tests began, the area had to be cleared. Most Indigenous people found in various sweeps by military personnel were directed to a mission at Yalata, far to the south on the Great Australian Bight (a community later ravaged by social problems). Ample evidence suggests, however, that individuals and small groups walked across the lands after the tests began. Those found from time to time in contaminated areas were given showers and driven away in trucks or cars, often suffering severe car sickness since they had never before travelled in a vehicle. In most cases, however, official ground patrols of the Maralinga range and adjacent areas simply turned Aboriginal people away, directing them to leave the test site. The locals who had lived around the Maralinga lands were scattered north and south, some never to return, others eventually to find their way to a small settlement at Oak Valley, 160 kilometres northwest of Maralinga.

Social problems have dogged the people of the Maralinga lands since the time of the tests. Three decades passed before they were compensated with a $13.5 million settlement from the Australian Government in November 1994. The people now hold the title to these lands under the *Maralinga Tjarutja Land Rights Act 1984*. After a partial handover in 1984, the final part of the Maralinga lands was officially handed back to the Maralinga Tjarutja people on 18 December 2009. These days, the only permanent settlement is Oak Valley, where about 90 people live.

The Maralinga Tjarutja people own the area that encompasses the Maralinga test site as well as Oak Valley, Ooldea and Emu Field; together these are now the Maralinga Tjarutja lands. These lands cover 105 667 square kilometres (or approximately 11 per cent of the land area of South Australia). The Anangu Pitjantjatjara Yankunytjatjara lands directly to the north encompass many of the Indigenous settlements affected by the British tests, including Wallatinna, Marla and Ernabella. The Western Desert peoples inhabit these lands, and the language usually spoken there is Pitjantjatjara. All these peoples have shocking family tales to tell of what happened when the men arrived with their planes and tanks and atomic weapons.

Ooldea Soak, 40 kilometres south of Maralinga, is one of the focal points of the Maralinga Indigenous saga. Ooldea was also known as Yuldi, Yutulynga and Yooldool in the various dialects of the diverse people who congregated there. According to scholar Odette Mazel, the name Ooldea means 'the meeting place where there is much water'. The Ooldea Reserve covered about 1500 square kilometres. The plentiful water supply made it an important meeting place and ceremonial ground for Western Desert people. When the transcontinental railway was built between 1912 and 1917, the influx of Aboriginal people to the area accelerated, as they were attracted by the capacity to trade with railway workers. Among the items traded were dingo scalps. The advent of the railway did much to change the old patterns of life.

The area is strongly associated with the Irishwoman Daisy Bates, an extraordinary character who established herself at Ooldea in 1918 and communed with the Aboriginal people there for 16 years. Her settlement was not strictly a mission, but she did provide the locals with food and clothes, while she studied their culture and made copious notes of her observations. During her time at Ooldea she was awarded the CBE and welcomed visits by royalty three times. She also befriended the writer Ernestine Hill, who made her famous by helping her to write a series of autobiographical features that appeared in several newspapers, some later published in *The Passing of the Aborigines*. Bates was a strict segregationist who opposed intermarrying and was profoundly pessimistic about the future of the Aboriginal people. She was never really accepted by the anthropological community, and her work fell into disrepute. The *Australian Dictionary of Biography* quotes a secretary who knew her briefly at the end of her life describing her as 'an imperialist, an awful snob ... a grand old lady'.

After Bates left Ooldea in the mid-1930s, the settlement was taken over by an evangelical missionary group called the United

Aborigines Mission. This group, dedicated to converting Aboriginal people to its particular brand of Christianity and attempting to eliminate traditional customs and beliefs, maintained the Ooldea Mission until 1952 when they chose to close it down. The water in the soak was drying up and soon there would be none left. Mazel stated that the activities of the mission represented the 'first active measures taken to interfere with aboriginal social orders', an ongoing process that led to considerable sorrow. Quite a bit more disruption and sorrow was to come.

In 1952, the order came from Minister for Supply Howard Beale that all Aboriginal people based at Ooldea would have to move to Yalata, about 120 kilometres south, in preparation for the start of British nuclear tests at Emu Field. Yalata was part of an old sheep station purchased by the South Australian government. It became a Lutheran-run mission for Aborigines forced to leave Ooldea. Many people from the Maralinga lands are there still. The 1984 Kerr Report found that

> the closure of the mission at Ooldea Soak in June 1952 in order to remove several hundred Aborigines further south to Yalata mission ... was of considerable anthropological significance. This was because rapid dislocation from their homelands caused confusion and distress among the Ooldea people which is held to be a major reason for the depressed and unhealthy state of the contemporary Yalata community – and is reflected also in other settlements containing dislocated Aborigines, for instance Cundalee [sic], Gerard and Ernabella.

The Aboriginal people of the Maralinga lands had complex ties to this arid place. Numerous groups of about 25 men, women and children constantly traversed the territory, encountering each other and reforming into new groups. At important ceremonial occasions, smaller groups also merged together for short times. As Europeans started to infiltrate the area in the 19th and early 20th centuries, some new patterns of Aboriginal activity began to develop. The opening up of pastoral stations brought the prospect of paid work for the first time, and the transcontinental railway offered opportunities for trade. Also, new movements of people were detected from the Warburton Ranges and the Gibson Desert towards Laverton, Mt Margaret, Kalgoorlie and Wiluna. People started travelling from Oodnadatta to Granite Downs too. In general, the early part of the 20th century saw considerable change in the way traditional owners moved around the land. The arrival of Europeans disrupted aeons-old movement patterns.

When the enormous Woomera Prohibited Area was declared in 1946, there were perhaps not as many Indigenous people pursuing traditional ways of life on the territory covered by the range as there had been 50 years before. But they had not ceased. The Royal Commission report found, 'The country was still used for hunting and gathering, for temporary settlements, for caretakership and spiritual renewal, and for traverse by people who moved from locations to other areas within and outside what became the prohibited areas'.

The Woomera project potentially affected far more people than that at Maralinga, since it involved firing rockets hundreds of kilometres across the centre of Australia. In February 1947 a meeting brought together various state and federal government authorities concerned with Aboriginal affairs, along with the anthropologists Professor AP Elkin, of Sydney University, and Dr Donald Thomson, and a Scottish medical doctor and Aboriginal rights campaigner, Charles Duguid. Duguid had campaigned against the land being used for rocket testing since 1946 and was a strong advocate for the Indigenous peoples. The meeting discussed two main issues: the physical danger from falling or exploding missiles and the acceleration of 'the de-tribalisation process in an uncontrolled and destructive fashion'. Duguid and Thomson, in particular, were concerned that testing military weaponry in the area would fatally disrupt the traditions of the Aboriginal inhabitants. Duguid fought hard against the proposal and resigned from the Aborigines Protection Board when he was unsuccessful. The upshot from that meeting was the creation of the position of native patrol officer in the federal Department of Supply.

Walter MacDougall was appointed, initially temporarily, as the first native patrol officer, beginning a legendary career. Tall, thin, gingery and very, very white-skinned, he was based out of Woomera. He carried the scars of an outback life – he was missing his right thumb and forefinger after accidentally blowing them off with his Winchester rifle out bush. He began work without even a designated vehicle and had to make increasingly long trips (sometimes over 6400 kilometres) alone. He was intelligent, empathetic and dedicated, and wrote long and impassioned reports and letters about his work. His was one of the few voices raised in support of the Aborigines during the rocket and atomic bomb tests. Government officials and scientists working on the tests saw Indigenous people as no more than a slight inconvenience.

MacDougall's particular way of interacting with Aborigines looks paternalistic today and shows signs of what we would call tough love. However, there seems little doubt that he cared deeply about the plight of the people being displaced by the atomic tests and did everything in his meagre power to try to curtail the harm done. He advised them about safe places to move to, adjudicated on disputes, hired some Indigenous people for pay and food, administered medical help when able, ferried people to doctors for more serious conditions and generally kept an eye on the welfare of the Indigenous groups who passed through the Maralinga lands. Most of the medical emergencies he encountered involved spear wounds, poisoned limbs, split heads, burns and pneumonia. His reports provide a candid and often wry glimpse into the lives of people affected by the bomb tests.

The territory affected was huge. Its outer extremes stretched across to Kalgoorlie in Western Australia, Port Augusta in South Australia and Alice Springs and the Canning Stock Route in the Northern Territory, an area referred to at the time as Western

Central Australia, the Central Aboriginal Reserve or the Western Reserve. The areas under threat were centred on Warburton and Ernabella missions, at the eastern end of the Musgrave Ranges not far from the Northern Territory border. The northern limit of the Maralinga Prohibited Area passed through the southern portion of the Central Aboriginal Reserve. Howard Beale claimed the area was 'so arid and waterless' that Aboriginal people did not use it. That was simply untrue.

MacDougall was a remarkable part of the story, and his compassion for the Indigenous people (even if tempered by condescension) made him incongruous. He was appointed a protector of Aborigines by South Australia in November 1947, after the Woomera tests began but well before the British atomic tests were planned for Australia. He was later awarded this same authority by the government in Western Australia. These appointments enabled MacDougall to enter or remain within the boundaries of any Aboriginal reserve or institution, in either state. He travelled extensively throughout the vast territory affected by the Maralinga and Woomera tests – a territory that eventually grew to around 800 000 square kilometres – and got to know the people well.

In a sense, he had a hopeless task. There was really nothing Mac-Dougall could do to save the inhabitants of the area from the vast and rapidly moving British juggernaut. It would roll over anyone in its way, especially people who were unrepresented in Australian society. That he took on the doomed job, and did it for years to the best of his ability in a harsh country, is a measure of his passion and dedication.

In 1954, the permanent test site had been newly named, surveyed and prepared for what was to come as MacDougall set out his views on the unfolding tragedy of the Aborigines of the Maralinga lands:

> Contact with white men has so far resulted in degeneration
> of the aborigines ... Because of their own socialistic way of

life, the generosity of their friends and friends' employers, and
government rations, they inevitably adopt the routine of moving
from Station to Station for free food. The result is laziness,
uselessness and loss of self-respect. They neither hunt nor work
for their food and the evils of unemployment of able-bodied
men are never brought home to them.

MacDougall drew a sharp distinction between the tribal Aborigines,
whom he mostly admired and whose lifestyle he considered to be
healthy and in tune with their environment, and the 'fringe natives',
who had been corrupted by the presence of Europeans and who, in
his view, had a lesser moral character.

MacDougall's first nuclear weapons–related survey of January
1953, ahead of the two Totem shots in October that year, found
400 Aboriginal people who made their 'headquarters' either at
Ernabella and Kenmore Park or Everard Park, Granite Downs and
Cullens. According to MacDougall, 'These people live largely off
the land, their only other source of supplies being bounty from
dingo scalps and government rations. Their ceremonial life is still
very active'. MacDougall proposed a full survey, to determine
exactly who was there and to what extent weapons testing would
affect their way of life. He asked for a Land Rover with trailer, fuel,
tools, camping gear, rations (including a tin of dripping), a rifle, a
camera, a compass and a cash advance of £5. The survey would
cover 3200 kilometres and take four or five weeks, visiting all the
Indigenous camps in the Everard and Musgrave ranges.

The trip proved quite an adventure. MacDougall encountered
some unexpected setbacks, due mainly to a lack of co-operation.

Whilst I have known most of these people for a long time, I
have not had an opportunity to discuss with them their secret
life, their relationships, their spiritual home country nor their
water supplies or camping and hunting areas. These people are
reluctant to discuss such important aspects of their tribal life

with just anybody and it was not until they fully realised that I
knew most of their secret life anyway and that I was not to be
put off with any cock and bull story or half truths, that I was
able to obtain the information that I required. I am afraid that
I was led upon arrival upon several wild goose chases at the
beginning.

In September 1953, a few weeks before the first Totem test at
Emu Field, MacDougall conducted a patrol of the area. 'The particular object of the patrol was to see where the 172 Jangkuntjara [*sic*]
Tribe people were situated after the annual [dingo] pup [hunting]
season exodus, and if necessary move them out of the prohibited
areas.' His first contact was at Wallatinna. From there he travelled
on a long and complicated path to Roxby Downs, Parakylia, Mt
Eba, Coober Pedy, Mabel Creek, Mt Willoughby, Wintinna and
Welbourne Hill. MacDougall was well acquainted with the rhythms
of the Indigenous seasons and knew more or less where he could
find tribal people at various times of the year. He recorded that 'on
Sunday 27 Sep I reported to the [Woomera] Site per radio that all
aborigines were accounted for and that I could return to Woomera
via Kingoonya'.

In December 1953, MacDougall was on the road again, looking
for older tribesmen previously based at Ooldea Soak to confer with.
He believed that the increased activity by Europeans around Emu
Field and Maralinga would be attracting them to the area. 'I believe
that a party with camels have been to Tietkens Well Area in the last
few days obviously attracted by signs of whiteman activity.' He constantly fretted, with good reason, that the more the white men made
themselves apparent, the more the lives of the Indigenous people
would be derailed.

MacDougall suggested a new trip in January 1954, to get a better
understanding of what the Indigenous people were doing and how
to deal with it:

The first move is to discover just what is likely to attract them. Ceremonial grounds, hunting conditions, water supplies etc. from their point of view. Secondly, everyday life conditions in their new area, also availability of new suitable ceremonial areas and to encourage the establishment of such areas.

MacDougall began this patrol on 13 January at Ooldea. The next day he encountered 300 Aborigines at Monburu tank, where 'Pastor Strelen was in charge'. He selected one of them, a young man named Sonny Williams, to accompany him on the rest of the trek, which took in three sacred sites. MacDougall's aim was to ensure that any sacred objects were removed from the sites, thus removing an attraction for the Aboriginal people to visit. He held the mistaken belief that moving sacred objects such as 'totem poles' would mean the sites would lose their sacred status. He found that sacred objects had indeed been removed: 'All significance lost'. (Possibly Indigenous people removed these when they were forced to Yalata, though MacDougall's notes were unclear about this.) Many years later, the Royal Commission examined this point and found that:

MacDougall's basic assumption was incorrect: removing sacred objects did not change Ooldea's status as a birth, death and dreaming site. Nor could this overcome the problem (for MacDougall) of people wanting to use Ooldea as a stepping-off point for sites to the north and west, or of people wanting to visit Ooldea from the north.

However, in 1954, MacDougall held that view that young people would not assign the same significance to the sites as the old people had, and that over a relatively short time the meaning of these sites, lost to weapons testing, would be lost to future generations too.

Secret life significance has ended mainly due to the lack of interest shown by the young people and the opposition to it by Missionaries. Owing to the fact that there are many of their relatives buried at Ooldea and that it is the actual birth place of many of them, there is a strong sentimental attachment. This will naturally die out in time.

Establishing Maralinga meant eliminating the Aboriginal reserve at Ooldea. MacDougall was told that this had been achieved bureaucratically in a letter from the secretary of the South Australian Aborigines Department on 14 January 1955. It said bluntly:

> Please note that the whole of the Reserve for aborigines, being Section 263 ... has been abolished. The area surrounding the Ooldea Soak is therefore not now an aboriginal reserve. It would be appreciated if you would remove any aborigines from this area when journeying through the district.

With the stroke of a pen, the lives of thousands were changed forever. MacDougall concluded that Ooldea was of no further use to the Indigenous inhabitants because their totems had been removed: 'No hardship would result from the withdrawal of the 900 square mile reserve provided some water supply is available at the Railway Siding'. There would be no impediment to nuclear testing from the traditional owners.

Many life-changing decisions were being made at this time. In July 1955 the AWRE decided, and the newly constituted AWTSC agreed, that they would build a meteorological station on 20 hectares in the Rawlinson Range to the northwest of the test site, just over the border into Western Australia. The station, 1500 kilometres from Maralinga, was to be called Giles, after an early explorer, and would assist in forecasting weather conditions before, during and after the Maralinga tests. In particular, the Giles station had to track the movements of air way up in the stratosphere, the currents that would affect where the atomic cloud ended up. Len Beadell and his team were brought in to grade a track from Mulga Park to Giles.

MacDougall had not yet been appointed a protector in Western Australia, so he had no jurisdiction there, but he was concerned about the effect of white people in a place frequented by Western Desert people who had had little contact with the West. When MacDougall threatened to make his concerns public via the Adelaide media, he was swiftly and brutally pulled into line. As Alan Butement made clear, the affairs of 'natives' were not to be placed ahead of the British Commonwealth. MacDougall was silenced and was only grudgingly allowed to participate in a September 1955 patrol to the Rawlinson Ranges with the Weapons Research Establishment's senior range reconnaissance officer at Woomera, TR Nossiter, in preparation for the construction of the Giles station. By then, though, the decision to build Giles had already been taken. Nothing MacDougall said could stop it.

Butement also complained in June 1956 that MacDougall was providing food to Indigenous people who had congregated at Giles, 'which action is hardly likely to ensure their early departure'. As the first major trials at Maralinga were only months away, this was a significant irritant. For Butement the presence of Aboriginal people was not the only problem: 'I believe that the natives are from time to time putting hazards on the jeep tracks in the form of spikes which at night time might cause a serious accident'.

The weather station at Giles was disastrous for the Aboriginal people, according to MacDougall's colleague Robert Macaulay, who was appointed as a second native patrol officer in 1956. He wrote in 1960:

> Giles Weather Station has had a marked effect on the Rawlinson
> Natives, increasing the number of wants they are unable to
> satisfy. Unless employment is made available soon, it could be
> said that the Weapons Research Establishment has not accepted
> the responsibility incurred in establishing a weather station in the
> Reserves.

MacDougall's report on one of his endless trips around the soaks and camps of the Woomera area provides a cameo of life in that harsh environment. During his patrol of July 1955 he followed up a message he had received that 'two women and an old fella' were at a water hole called Warrapin. He discovered two old women but no man.

> They were very frightened and had not tasted white man's food. They liked oatmeal, rice, damper, tea and sugar. They were very doubtful about anything from a tin. Their possessions consisted of the following: 1 digging stick each, 2 wooden dishes, each full of grass seed, 1 upper mill stone, 1 piece of iron rod, sharpened at one end, and one white dingo.

MacDougall concluded that the bigger group had left the two old women behind 'to fend for themselves for the rest of their lives'. He gleaned this from talking to the women and to the man who had found them, Frank the native guide. MacDougall and Frank talked the women into accompanying them to the Cundeelee Mission, run by the Australian Evangelical Mission, 40 kilometres north of Zanthus, 'where they could be cared for in their old age', but when it came time to go the women decided that they couldn't leave their land. MacDougall left blankets, flour, rice, tea and sugar with them and went on his way. He had tried to get them to go to Cundeelee even though he did not think much of the place: 'I was not favourably impressed by Cundeelee Mission. The Missionaries by their attitude suggested that they believed that the world owed the natives a free and easy living, an attitude that suggested that the natives should be given everything they want'. Cundeelee Mission, which had been under the control of the evangelical mission since 1950, held a sandalwood licence, and Aboriginal people were paid £26 per tonne of wood they harvested. Some handicrafts were made there too, in a sort of cottage industry. Part of the self-imposed role of the mission was to entice 'bush natives' away from their traditional lifestyles and evangelise them into its particular brand of Christianity.

In June 1956, plans were in hand to test for fallout throughout the Maralinga lands. The AWTSC, headed at that time by Professor Martin, recommended that fallout monitoring kits, essentially comprising sticky paper to catch any swirling particles, be set up around the areas known to be inhabited by Aboriginal people. As the fallout from the Maralinga tests was likely to be to the northeast of the site, the AWTSC recommended that the sticky papers be placed at Ingomar Homestead, Mabel Creek Homestead, a shed halfway between Mabel Creek and Mt Willoughby, Mt Willoughby itself, Welbourne Hills, Granite Downs, Echo Hill and Ernabella. The sticky papers were to be changed daily during each test period, plus a few days before and after.

As the date for the opening of the new Maralinga range approached, MacDougall was busier than ever. In mid-1956, he undertook a patrol of a large area that encompassed not just the weapons range but also a new area set aside for the South West Mining Company. In his report, he mused about what was going on:

> They, still in the Stone Age – hunters and gatherers with a
> code of laws and social customs effective only whilst they are
> segregated, and with harsh penalties applied – cannot continue
> to exist unchanged within our civilisation with its amazingly
> rapid scientific development, but they are human beings and
> must be considered as such.

MacDougall set out his estimate of how many Aboriginal people lived in the area affected by the tests. As at November 1955, there were 1000 people and their numbers were increasing. Research-based calculations suggested, he said, that the numbers were expected to double in 20 years. He broke these numbers down into various areas, thus: Everard Park 200, Musgrave Ranges 350, Warburton Ranges 350 and Rawlinson Ranges 100.

The detrimental effect of the opening up of this area depends
upon the policy decided upon and the extent to which the
policy is effectively policed. The policy of controlled contacts
as provided for at present has been hopelessly broken down.
Segregation is now impossible.

MacDougall made the point that while around 1000 Aboriginal
people lived in the area, only 50 per cent of them lived off the land:
'One thousand natives will be more or less affected by the establishment of the Range in this area. None of these were completely
uninfluenced by contacts that have been made, although some have
never seen white men'. He was greatly opposed to giving Aborigines handouts and scathing of Europeans who thought otherwise.

Because of the complete inability of many of the personnel to
understand the different way of life of the aborigines, it makes
it difficult for them to understand that their normal notions
and reactions are detrimental to the welfare of the aborigines;
e.g. 'I know that that man is hungry because I have seen him sit
there all day and he has had nothing to eat. I cannot harm him
by giving him something to eat'. To explain that the man would
not be sitting down all day looking hungry unless he knew that
he would be freely given better food and water than he or his
ancestors ever had before, makes no impression.

In this letter, MacDougall recommended that the problem of the
Aborigines in the area 'be treated as one of great national importance'. He also advocated for a move that took more than a decade
to eventuate: 'that steps be taken to unify [Indigenous] policies, laws
and regulations throughout the Commonwealth'.

Beale took a sanguine approach to the Indigenous issue when
planning for the British atomic weapons testing project to descend
on the Australian desert. The government was perfectly happy with
William Penney's advice that X300, now named Maralinga, was the

place to establish a permanent test site. In a top-secret Cabinet briefing document, Beale falsely claimed that if Maralinga was chosen they could revoke the existing Aboriginal reserve at Ooldea without difficulty as Aborigines had not used the area for some years.

Robert Macaulay, MacDougall's colleague, was only 23 and fresh out of the University of Sydney when he was appointed as the second native patrol officer. He had no experience in the outback and in fact had rarely been outside Sydney. Initially based at Giles, his job was to ward off Aboriginal people and report on their whereabouts to the test authorities. He was woefully ill equipped, not only in life experience but also in gear – he had no car or radio when he started. He eventually borrowed a car and made his first trip south to the Ernabella Mission on 12 September 1956, two days after the first Buffalo shot was scheduled. He did not have a radio, though, so he was largely out of contact with his masters.

In the event, Buffalo was delayed until 27 September, but even so preparing the local Aborigines for the first Maralinga major trial was impossible due to the short time and huge distances. On the day of that test, Macaulay sent a cable from Giles to Woomera, saying:

> Unable to satisfy myself no natives south of mentioned line.
> Have not been there. Unable to penetrate without own vehicle
> and radio. Hear none Mt. Harriot area. Regret unable to signal
> daily. Unaware I could use the flying doctor system. Shall remain
> Giles until vehicle and instructions arrive.

Even though there were now two men, MacDougall and Macaulay, the task was getting more difficult. As doctor and activist Charles Duguid said in a 1957 speech, 'It is an utter impossibility for two men efficiently to patrol such an area, particularly as tribal aborigines are always on the move and can keep out of sight if they wish'. And so it proved to be – the territory they were expected to cover was so vast that it was not humanly possible. They did what they could, but it was not enough. Duguid's warnings were stern:

The British Government, the Federal Government of Australia,
and the Governments of South Australia and Western Australia
must all join to ensure the future development of the people
of the Central Reserve whose territory they have invaded.
But they must act quickly or it will be too late to redeem a
situation fraught with tragedy for the natives and shame to
ourselves.

MacDougall was an opinionated individual and a vocal critic of
many decisions made on behalf of the Indigenous inhabitants of the
Maralinga lands. He opposed the policy of keeping tribal Aborigines segregated from white people, 'since it is obviously impossible
to keep tribal Aborigines segregated for ever'. However, when the
policy changed in 1956, he expressed his anger forcefully:

I understand that the policy has now been changed though
oddly enough I, as Dept. of Supply Native Patrol Officer, was
not informed nor has it been promulgated in any way. The
new policy appears to be a third and disastrous alternative
whereby contacts are made by completely unqualified persons
and no provision is made to train the Aboriginals to fit into the
twentieth century. The result is certain to be a degeneration
from self-respecting tribal communities to pathetic and useless
parasites – it has happened so often before that surely we
Australians must have learnt our lesson.

For all his paternalism, MacDougall knew what was at stake: 'The
country under discussion belongs to the tribe and is recognised as
such by other tribes. However, we propose to take it away from
them and give nothing in return – we might as well declare war on
them and make a job of it'.

After the 1956 Buffalo tests, the *West Australian* newspaper ran
several stories about the effects on Indigenous people. The deputy
leader of the Opposition Arthur Calwell wrote to Beale asking if

press reports about Aboriginal children being separated from their parents were true. HJ Brown, the Weapons Research Establishment controller at Woomera, sent a cable briefing the department so that they could reply to Calwell:

> There was no separation of Aboriginal parents and children anywhere in Australia as a result of the atomic trials. The only arrangements made respecting Aborigines was to keep track of their movements and to maintain information on their whereabouts. If necessary their movements were controlled to ensure that they did not enter danger areas but this was hardly necessary as they appear to be aware of the necessity for keeping away from the areas involved.

The message said that the issue of the atomic trials was probably being conflated with the practice of the Warburton Mission 'to take children from the Aborigines and endeavour to keep them at the Mission station for training. This is a policy of the Mission which is creating some criticism but has nothing to do with atomic tests and has been in operation for some time'.

The native patrol officers did not necessarily interact happily with that other great bushman of the region, Len Beadell. On a patrol in March 1957, Macaulay encountered Beadell driving his Land Rover, with an Aboriginal man as his passenger, looking for a landmark of Giles the explorer. Macaulay was furious. 'Beadell thought his own need justified the breach of regulations. I did not.' Macaulay notified TR Nossiter at Woomera. 'I consider that not only was Beadell's action a definite breach of the Controller's instructions, but that it was detrimental to the native way of life, and more important, detrimental to the contact situation which I am attempting to control.'

The AWTSC meeting of 19 July 1957 in Melbourne, chaired by Ernest Titterton, reviewed plans for the final major trials series at Maralinga and responded to a letter from the secretary of the

Aborigines Board in Adelaide. 'Similar facilities for moving aborigines are required for the Antler tests as were available during the Buffalo tests', the minutes laconically related, as though making arrangements for the movement of cattle.

MacDougall's report of October 1957 told of his efforts to clear the area of Indigenous people before the Antler trials the previous month. During these trials, MacDougall based himself 'at a strategic point upon the Emu–Giles Road'. Macaulay went to the Everard Ranges – Officer Creek area to check on the whereabouts of the 'Jankantjara' (probably Yankunytjatjara) people who were known to be hunting dingoes in the area at the time, while at Betty's Well in the northwest corner of the Everard Range, 150 Jankantjara people were concluding an initiation ceremony. Both MacDougall and Macaulay hired local Indigenous people to assist them. William was engaged by MacDougall on 'a food and transport with time to hunt dingoes basis', while Tom Dodd, 'an old halfcaste', was offered £2 per week to travel with Macaulay.

MacDougall found 27 Aborigines hunting dingoes in the Mt Lindsay area. These were Ernabella Mission people from up north. They had been briefed on the forthcoming Maralinga tests and knew who to contact if they experienced any problems, according to MacDougall. On his travels from Coffin Hill to Rawlinson Range, Mt Davies and Ernabella, he encountered a small number of Indigenous people. He then joined forces with Macaulay at Shirley Well and travelled around further, again discovering few Aborigines. MacDougall wrote in his report, 'It is comparatively easy to ensure that a definite area is free of natives and definite information can be checked and forwarded. It is unfortunate that the dingo pup season coincides with Maralinga tests but the natives are quite content to keep out of the areas when told to do so'.

The incident that, decades later, became the most famous concerning Aboriginal people during the British nuclear tests personalised for many the folly of conducting nuclear tests on land where people lived. On 15 May 1957, the Milpuddie family – Charlie

(Tjanyindi), Edie, Henry (Kantjari) and Rosie (Milpadi), noted in the report as 'Father, Mother and two Picaninnies', along with their four hunting dogs – were found in a very inconvenient place. They had camped overnight alongside Maralinga's only atomic bomb crater. A party of Royal Australian Engineers led by Captain Rudi Marqueur spotted Charlie at 9.15 am, as he walked from the crater to the health physics caravan at Pom Pom. Captain Marqueur noticed that the man gave hand signals to indicate he wanted water. After having a drink, Charlie led Marqueur back to the camp where his wife, children and dogs were waiting. The family had 12 dingo pelts gathered during their journey, almost certainly collected for the bounty payable. The huge crater had been created seven months earlier at the Marcoo test site as part of Operation Buffalo, by the only atomic bomb detonated at ground level. The area had been classified on site as dirty, meaning contaminated.

The acting security officer for Maralinga, B White, met the family at the caravan. The report from Sergeant Frank Smith noted that 'both mother and daughter were very shy as regards any approach in the early stages'. The health physics officer on site, Harry Turner, was initially quoted as saying at the time that they were all free of contamination, although this was not true. The boy, Henry, thought to be about 11, had contamination on his hair and body, and the other family members were not thoroughly checked. Turner's own report provided more detailed information. He said that they were 'monitored head to foot' and the only trace of radioactivity was found on the boy.

> The boy was then persuaded to shower in the caravan. He
> was thoroughly washed by Mr. D SMALL who paid particular
> attention to the boy's hair. At the conclusion of the operation,
> the boy was a new person and was so obviously pleased at the
> result that it was not difficult to persuade the father to shower.
> The father then washed his daughter. The mother was content

with just washing her hair. Altogether the process of monitoring and washing was accomplished surprisingly well, considering the circumstances.

Turner then handed them over to security, who 'evacuated them from the area'. The Milpuddie story came to be known as the Pom Pom incident after the location where the father had been found.

The family must have arrived at Marcoo sometime after 6 pm the night before, when the health physics caravan had closed for the day. They had approached from the northwest, walking across about a mile of land contaminated by fallout, which Turner asserted 'would not adhere to their feet or bodies'. The campsite was also contaminated, although Turner said that a 16-hour stay at the site would expose them to

> only about 2% of the weekly dose that is permissible every week
> throughout a lifetime, and about 0.025% of the dose required
> for clinical detection. The inhalation hazard was completely
> negligible. The contamination on the hair of the boy was less
> than the accepted tolerance value ... Therefore there is no
> possibility that any of the family could have experienced any
> radiation injury.

The family were Spinifex people from the Ernabella Mission and had travelled to Ooldea to visit relatives, not knowing that it had been closed down. They were shipped to Yalata and placed into the care of Pastor Temme of the Lutheran Mission. Edie was pregnant at the time, and soon after she miscarried. They took their dogs with them, but when Howard Beale found this out he issued a direct order to the range commander that all four dogs be shot. This was done in front of the family. Over the next few years Edie suffered several more miscarriages. She was interviewed in 1985 when the Royal Commission went to the outback and sat in the dust with the Indigenous owners. She eventually revealed the sorrow of her miscarriages to Jim McClelland, and he featured the story in his report.

On 3 December 1957, Beale was again quizzed in parliament about reports of harm to Aboriginal people from the Buffalo and Antler test series, and HJ Brown again had to come up with acceptable answers quickly. John Moroney sent an urgent cable to him at Woomera, recounting the questions:

> (1) Did a mystery disease of epidemic proportions a few months ago result in a number of deaths amongst Aboriginal children at the Ernabella Mission Station in South Australia? (2) Is it a fact that in certain quarters the deaths of these children were attributed to the effects of radioactive fallout from bomb tests? (3) If he has not already done so, will he have this report investigated immediately and make information available as soon as it comes to hand?

Brown's response noted that 'a disease of epidemic proportions did occur at Ernabella during March to June 1957, resulting in the deaths of 20 children and 2 adults'. He denied it was a mystery disease, 'although soon after its outbreak a pathological investigation carried out at Alice Springs apparently did not disclose the cause of death'. He said that 'three children were removed from Ernabella to the Children's Hospital Adelaide where subsequently one died and the result of post mortem indicated large fatty liver and malnutrition, and infection of both mastoids. At least one of these children was admitted with a history of measles'. Further investigation, he said, had indicated an outbreak of measles. Medical scientists had not investigated 'the possible effect of radio active fall out', but they had 'since stated that there was no reason to suspect any other cause for the deaths than measles'. Brown also pointed out that an influenza outbreak at Ernabella had killed some people. He rejected speculation in the British Medical Association journal *The Lancet* that had drawn a connection between measles and radioactivity, piquing some interest and questions from Charles Duguid. Even if the article were true, he said, 'Aborigines have not been subjected to radio activity'.

As we have seen, after the major bomb trials ended with Operation Antler in 1957, the activities at Maralinga were more secret than ever. The more dangerous minor trials meant that Aboriginal people still had to be kept away, although it was harder to explain why. A patrol report by Macaulay in October 1963 mentioned ongoing issues:

> The [Maralinga] Range Commander had recently travelled
> over the outer perimeter roads and had met an Aboriginal family
> in the Prohibited Zone. This had led him to some appreciation
> of the problems and the delicate handling required in the early
> stages of contact between whites and nomadic Aborigines,
> especially in such a political context for Aborigines and the
> Maralinga Prohibited Zone.

On 1 May 1963, Jim Cavanagh, Labor senator for South Australia (and later federal minister for Aboriginal Affairs in the Whitlam government), asked Beale's representative in the Senate about the Aboriginal interaction with the Maralinga site. The Department of Supply prepared a briefing paper for the minister to help answer these questions. For the question 'Have experiments with nuclear explosions been conducted at Maralinga, South Australia?' the paper suggested that the response '"yes", qualified by "not since 1957", is strictly true'. Actually, this was not strictly true as the nuclear experiments of the Vixen B series continued until April 1963.

Cavanagh also asked two questions as to whether Department of Supply personnel and vehicles were to keep native Aboriginals off the area, and whether Aboriginals were kept off the area at the time of such experiments. The briefing paper disputed the implication

that we had to take positive action to <u>keep</u> aborigines off the
area. The facts are that we conducted land and air surveys, and
found no natives in the area at the time in question. We did
not have to keep them off. However, one group of natives was
encountered on walkabout <u>outside</u> the area, and they were
transported (with their willing consent) to their destination.

Even had this action not be taken, the briefing paper asserted, there
was no reason to believe 'they would have been in the area at the
time of the explosions'.

A fourth question asked if the government had 'any knowledge
of harmful effects on natives as a result of these explosions'. The
briefing document answered by saying:

There have been attempts to blame the tests for various ills
suffered by the natives, but after investigation we have been
unable to find any foundation for such claims. One instance
of penetration into the area subsequent to the trials is known,
but the natives concerned were removed from the area, and
were given a thorough decontamination and examination. They
were taken to the Aboriginal Mission at Ooldea [actually, the
Milpuddies went to Yalata], where they remained for some time,
but no detectable effects were observed.

One of the most perplexing episodes in this history is the
so-called black mist. This little-understood phenomenon occurred
after Totem 1 at Emu Field in October 1953. Many later commentators have wrongly associated it with Maralinga. The black mist
story has gone into local Indigenous folklore, but its first broad
public airing was in the Adelaide *Advertiser* on 3 May 1980. When
the Royal Commission investigated the Emu Field tests, it concluded that weather conditions for the first Totem blast had been
unsuitable and the test should not have proceeded.

The Indigenous people most prominently associated with the

black mist are Yami Lester and Lallie Lennon, both of whom experienced it and reported their experiences publicly, and to the Royal Commission. The allegation is that after the first bomb in the Totem series was detonated on 15 October 1953, an unpleasant greasy black cloud enveloped the land around Wallatinna and Mintabie and deposited material on the people in the vicinity. Uniformly, they reported vomiting, diarrhoea and skin conditions, as well as blindness (in the case of Yami Lester, who was a child at the time) and a number of deaths. Ernest Titterton called the allegations 'a scare campaign' and denied the possibility of a black mist. He said, 'If you investigate black mists you're going to get into an area where mystique is the central feature'. William Penney stated to the Royal Commission, 'I was not aware at the time of any of the alleged reports of "black mist"'.

Totem 1 was a 9.1-kilotonne atomic device detonated from a 30-metre-high steel tower at 7 am on 15 October 1953. The evidence is unclear exactly how long after the detonation the black mist rolled across the land, but some estimates say it was first seen about five hours later. Wallatinna is 173 kilometres from the Emu Field test site, and Mintabie just over 16 kilometres from Wallatinna. Scientific estimates have confirmed that the fallout cloud would have passed over the area about five hours after the detonation.

British scientists WT Roach and DG Ballis in evidence to the Royal Commission supported the possibility of a black mist from the Totem test, saying that all the reports 'had a measure of internal consistency about them'. Both scientists asserted that the conditions of firing at Totem could conceivably have delivered to the Indigenous people of Wallatinna a fallout cloud that had raced along near ground level. 'It would have been a strange and awesome sight to anyone beneath it. A fine "drizzle" of black particles would also have been noticed.' However, they did not think that the cloud would have caused health problems for anyone standing awestruck beneath it.

And that is the problem. While most people who have looked at this issue in any depth agree that the black mist occurred, disagreements about the harm it caused have never been resolved. But the compelling evidence of the Aboriginal people in the area is hard to ignore. The most famous of those affected, Yami Lester, said, 'Almost everyone at Wallatinna had something wrong with their eyes. And they still do ... I was one of those people, and later on I lost my sight and my life was changed forever'. One of the Aboriginal people who was a spokesperson for Aboriginal witnesses to the Royal Commission, Kanytji (also known as Kantji), said the cloud was 'very black', but reddish towards to the top, and it dimmed the sun. It produced a strange shadowing effect, seeming to give people multiple shadows. Kanytji said that the cloud deposited a moist black substance on the ground, like a bizarre kind of frost. Other eyewitnesses said that it smelled like a dead kangaroo or like liquid petroleum gas. Whatever it was, it was not healthy.

Despite Titterton's scepticism, the Royal Commission officially recognised the credibility of the black mist allegations but found insufficient evidence to say whether the phenomenon caused injury or illness. This finding still causes considerable distress to the Indigenous people who were present when the black mist rolled in.

More recently, scientists from ARPANSA have attempted to get to the bottom of the black mist, including Dr Geoff Williams, from the team who uncovered the radioactive contamination at Maralinga in 1984 and who has extensive knowledge of the British nuclear tests. In 2010, with his colleague Dr Richard O'Brien, he carried out a scientific appraisal of the black mist in response to *The Black Mist and Its Aftermath: Oral Histories by Lallie Lennon*, prepared by oral historian Michele Madigan in 2006 and 2009 and published with transcription and commentary by Paul Langley in February 2010. The scientists affirmed the strong evidence that the black mist incident happened but, again, were unable to say that it caused illness and injury. They pointed instead to the measles epidemics around the time of Totem 1 as more likely causes. They also raised the possibility that 'non-radioactive factors', such as chemical irritants in the mist, might have caused the reported skin conditions and allergic reactions.

The terrible harms caused to the Indigenous peoples of the Maralinga lands have been partly salved by the determination of the people themselves not to be defeated. One measure of their spirit is the establishment of a tourism venture at Maralinga owned by Maralinga Tjarutja people based at Oak Valley. Maralinga Tours now takes paying customers to the old test range. The audacity of this venture cannot but lift the spirits.

8

D-notices and media self-censorship

The press in both Britain and Australia, at least initially, did not probe at all into the political, scientific, moral, economic or any other aspect of the atomic project. They allowed themselves to be bound by a series of D-notices.

Robert Milliken, *No Conceivable Injury*, 1986.

When the media acquiesce, the very existence of censorship is unknown to citizens. In Australia, D-notices, used to censor the media, seldom receive publicity.

Sue Curry-Jansen and Brian Martin,
'Exposing and opposing censorship: backfire dynamics
in freedom-of-speech struggles', 2004.

Why bother to muzzle sheep?

Attributed (possibly incorrectly) to Ernest Bevin,
postwar UK foreign secretary, 1940s.

In the 1950s, Australian newspapers were popular, opinionated and dominated by legendary media barons, notably Ezra Norton, Keith Murdoch and Frank Packer. For the most part they focused on growth, politics and postwar prosperity. Media owners and editors were not used to covering scientific issues, and there was no imperative to do so. They filled their pages with the economic and population boom, and the red scares and paranoid preoccupations of the Cold War. The intricacies of nuclear weaponry were not at the top of the minds of newspaper people, even as the pall of mutually assured destruction descended on a nuclear-armed world.

Since Australia had no nuclear energy or weapons program of its own, the country and its media were several steps removed. And because they were not attuned to matters nuclear, the Australian media were easily controlled when it came to managing atomic weapons secrecy. Ignorance was helpful in this process. The British nuclear test authorities prepared the way well to manage the media and did so with almost unbelievable success throughout the entire test program.

Why were the Australian media so compliant to the secrecy requirements of the test authorities? At least part of the answer may be found in the top-secret agreements between government and media called D-notices, which encouraged media self-censorship. While their influence was relatively fleeting, D-notices in Australia had their greatest impact during the 1950s. The D-notice system established a formal co-operative relationship between the government and the media in the lead-up to, and during the first few years of, the British nuclear tests in Australia. This relationship set specific reporting ground rules – rules that for the most part the media seemed willing to obey.

The imposition of controls over the media, exercised by both the British and the Australian governments, arose from a long chain of circumstances. A vicious world war in recent memory. The rise of the Soviet Union, a major totalitarian state – now a superpower – whose postwar armaments and strength derived at least partially from the leaking of official Western secrets. To Australia's north, there was the 1949 communist revolution in China and in 1950, just before Clement Attlee asked Robert Menzies about atomic testing in Australia, North Korea invaded South Korea. When United Nations forces, mainly American, were called to defend the South fears grew that the Cold War might become hot at any time. Australian troops soon joined the action. All of these factors made Australia generally, and its government in particular, jittery and insecure – and ready to do whatever it took to buy postwar security.

In addition, a series of postwar scandals about supplying security information to the Soviet Union had implicated the Australian public service. Australia found itself in the uncomfortable position of needing to convince both the UK and the US that it could keep security secrets. Collaboration on national security issues between Australia and its two main allies depended upon making fundamental changes to the way Australia conducted itself, particularly with the new dynamics around nuclear weaponry and the arms race. In June 1948, relations with the US were ruptured when Washington suspended the flow of classified military information to Australia. Combined pressure from Westminster and Washington led to the establishment of ASIO as a domestic spy service in March 1949.

Later that year Menzies led the Liberal Party to its first term in government, ushering in a lengthy era of conservatism. Menzies, a former constitutional lawyer, was a fatherly figure, with his shock of white hair, his beetle-brow, his tall and well-built frame and his double-breasted suits. Revisionists have since portrayed his politics as opportunistic, relying upon Cold War fears to shore up his support, but during his stately, 16-year second prime ministership, mainstream Australia mostly viewed him as solid and safe.

The Liberal Party of Australia was formed in 1945. Menzies had previously been prime minister between 1939 and 1941 when he headed the soon-to-be-defunct United Australia Party. His short and unhappy two years as a wartime prime minister taught him valuable lessons. Keeping tight control of national security matters came naturally to him, and his distrust of the media had a long history. His second prime ministership came at a time of pervasive anti-communism in Western countries, set in motion in part by the aggressive pursuit of atomic weapons by the Soviet Union. Menzies was elected to his first postwar term on a national security platform and a pledge to outlaw the Communist Party. The government passed anti-communist legislation, but its bill was struck down by the High Court. A referendum intended to give the Australian parliament the power to ban the Communist Party was narrowly defeated in 1951.

Despite this failure, Menzies gained great political capital out of defence and security issues throughout his long tenure, which spanned seven general elections, until his retirement in January 1966. In his autobiography, Menzies wrote about tightening national security with the advent of ASIO, noting that his predecessor Ben Chifley 'laid down a rule, which I subsequently strictly observed, that ASIO must work in secret (since it was trying to counter an enemy who worked in secret), and that the details of its activities should not be exposed in Parliament or to the public at large'.

ASIO's role, roughly equivalent to that of Britain's MI5, was to deal with internal security issues. The Australian Secret Intelligence Service, a new (and for many years totally secret) external security organisation equating to Britain's MI6, was established in 1952, although the government did not acknowledge the service publicly until the 1980s. It remains subject to a (notional) D-notice to this day, despite some controversial breaches by the Australian media over the years.

Fear was in the air and the world seemed more dangerous than ever. The atomic arms race escalated, along with international political tensions. Britain had been testing long-range guided missiles since 1946 at Woomera in the South Australian desert, and, apart from the brief cessation in the late 1940s because of security scandals, the Woomera test range was used until 1980. In fact, Australia hosted a large proportion of the UK's postwar weapons testing program. Given this central involvement in the weapons testing activities of another nation, initiating an Australian D-notice system to manage media information now looks as inevitable as establishing a spy service.

The British Government needed Australia's geographic assets and its distance from the British electorate but was not convinced of Australia's soundness in managing security issues. These doubts were not helped by the Petrov spy scandal that began in April 1954, on the eve of a federal election, when Vladimir Petrov, a Canberra-based junior Soviet diplomat, defected to Australia. Petrov claimed there was a communist spy ring in Australia that included diplomats, journalists, academics and even Labor Party staff members. This quintessential Cold War saga was sparked by the disruption in the Soviet Union after the death in March 1953 of the cruel dictator Joseph Stalin. Petrov feared for his life if he returned home. His wife, Evdokia Petrova, taken forcibly by armed Soviet escorts and on her way back to Russia, was rescued by ASIO agents at Darwin airport on Menzies' direct orders. A famous photograph of her being manhandled by KGB operatives on the tarmac evokes the paranoid atmosphere of that time. The drama led to an Australian Royal Commission on Espionage and the severing of diplomatic relations with the USSR until 1959. In fractious and paranoid times, Menzies brought security issues to the top of the political agenda.

When its security record caused significant damage to its international relationships, Australia had some fence-mending to do. The Americans in particular were extremely wary of Australia's approach to security, and the British became hypersensitive about lax secret keeping, particularly as they developed their plans to test nuclear weaponry. In fact, evidence suggests that the British even used Australian security slackness as an excuse. An official from the CRO, the authority that liaised between the British Government and members of the Commonwealth, wrote in April 1952, 'By explaining to the Australians the [security] measures we consider satisfactory, we shall deprive ourselves of the easiest excuse for withholding from them information about atomic matters in the future'. The perception of slack Australian security could be valuable to the Brits.

Democratic governments have to tread carefully when they seek to manage the media. Media culture (if not the daily reality) adheres strongly to the idea of the fourth estate, which demands that journalists hold power structures to account. This idea arose during the Enlightenment and forms a backdrop to media in democratic countries. Limiting the media by legislation is politically unwise because the backlash can be brutal. How, then, to achieve control? In Australia, for a short while, as the insecurities and heightened patriotism of the Cold War played out, D-notices became the favoured mechanism. These new directives particularly influenced the coverage of the early atomic tests.

The D-notice system, guided by a secret committee that numbered senior media representatives, politicians, bureaucrats and military leaders among its ranks, set up a dynamic between the British nuclear test authorities, the Australian Government and the Australian media. It proved an effective way to get the media to report officially vetted information and to dissuade them from seeking other sources for their stories. The media were in effect 'trained' not to step into the realm of independent inquiry in relation to the British nuclear tests. The D-notice system was important in establishing this relationship.

The British system of D-notices, short for Defence notices, dates back to 1912. That system was set up following the proclamation of its famous *Official Secrets Act 1911* and has been used extensively since. It has never had any legal authority, since legislation aimed at media censorship would cause unproductive outrage among media organisations and the general public alike. Also, for the UK authorities D-notices were a way of ensuring 'prior restraint' – in other words, media self-censorship. Convincing media practitioners not to publish national security information was less hazardous and more effective than pursuing media outlets if and when they did so. Australia has never had an exact equivalent of the Official Secrets

Act, although the *Crimes Act 1914* does cover aspects of unauthorised disclosure of Commonwealth classified information.

The Official Secrets Act (which was updated in 1990) applied to all British test personnel, military and scientific, in Australia, while the Crimes Act covered Australian personnel. Both limited what participants could say about what they knew and saw. The British legislation was especially rigid, stipulating lengthy prison sentences for breaches. Many of the people who took part in the tests would have found this sufficiently threatening to silence them, and certainly to constrain them from talking to the media. The Australian Crimes Act was not as prescriptive, and the penalties for talking to journalists were less severe. However, the laws of both countries placed real restrictions on potential sources for journalists. But while that side of the equation – the potential sources – seemed under control, knowledgeable insiders could still surreptitiously disclose information to the media. The D-notice system was established to manage the other side of the equation, the media themselves.

As in the UK, D-notices in Australia operated without a legislative foundation. They began in the UK as a peacetime mechanism, less onerous than more restrictive wartime media controls, and this remains the case. In their heyday (and particularly during the Cold War), D-notices flattered media organisations by treating them as equals with as much a stake in patriotism and national honour as the government. D-notices also provided an orderly mechanism whereby media could publish agreed information on national security matters without risking litigation or being harried by security authorities generally. D-notices in Australia are largely unknown, even after they were publicly revealed in 1967. They never became an entrenched feature in Australia. One scholar, Pauline Sadler, believed this was partly because 'the print and electronic media in Australia never had specialist military correspondents of the calibre of those in the U.K., such as Chapman Pincher'. (Pincher, who lived to be 100, tested the limits of the British D-notice system of media self-censorship with a 1967 story on the interception of communications by UK security authorities. Ultimately no action was taken, though, because he had not actually broken the UK Official Secrets Act.)

The British Government had exerted pressure on the Chifley government to adopt a D-notice system in Australia when it decided to test missiles at Woomera. Chifley acted to control the coverage of the missile program by ensuring public statements about it could come from only himself or his Defence minister John Dedman, but he was reluctant to commit to a D-notice system. After several informal approaches, Edward Williams, the UK high commissioner, made a formal approach on 28 January 1947. A subsequent discussion between Chifley, Williams and Lieutenant-General JF Evetts, the British officer in charge of Woomera, led to some action on publicity arrangements for the Woomera project, but nothing regarding D-notices. An account of this discussion later prepared for Menzies noted that the British high commissioner had inquired whether D-notices could be extended to Australia since, like Canada, the government was 'in possession of much secret information supplied by the U.K.' and was 'responsible for trials and defence research based thereon'. He had also pointed out that in December 1949, the Australian Defence Committee had suggested 'that every endeavour should be made to obtain the co-operation of the Press in the adoption of a system of "D" notices'.

Despite the pressure being applied by the British, the Chifley government was unreceptive, possibly because of a perception that the D-notice committee itself might leak information. Security issues continued to dog the government as the Cold War worsened, but D-notices progressed no further until Menzies came to power.

One of the most painful episodes in the dying days of the Chifley government concerned the Council for Scientific and Industrial Research (CSIR), forerunner to the current Commonwealth Scientific and Industrial Research Organisation (CSIRO), Australia's leading scientific research agency. CSIR under its chair David Rivett became mired in controversy over its role in defence science and the adequacy of its security controls. Rivett famously gave a speech titled 'Science and responsibility' in 1947 saying CSIR should not be involved in any research that could not be openly published and discussed in scientific fora. A political and media debate ensued around the loyalty of CSIR's scientists and management in light of the allegations of communist infiltration. This ultimately led to legislation to create CSIRO, removing the capacity to carry out defence science. Defence science was shifted to an organisation within the Department of Defence now known as the Defence Science and Technology Group. Nuclear weaponry, its associated espionage and official secrets imposed new limits on what scientists could do and say.

For the British, plans to test nuclear weapons in Australia made the issue increasingly urgent, so the UK stepped up its pressure on the more amenable Menzies. The secretary of the Australian Department of Defence Sir Frederick Shedden – a long-time supporter of an Australian D-notice system – secured Menzies' agreement.

Menzies had previously approved mechanisms limiting the media. During his wartime prime ministership, he had placed the entire Australian media under the control of the director-general of information. The influential Australian newspaper proprietor Keith Murdoch, father of long-time News Corp head Rupert Murdoch, held this position briefly. (Wartime restrictions placed upon Australian newspapers were maintained – possibly even strengthened – by Menzies' Labor successor, John Curtin, straining the relationship between Curtin and the press.) Even by wartime standards many deemed this measure excessive. An editorial in the *Sydney Morning Herald* commented:

> The new regulations give the Director-General, subject only to
> the direction of the Minister – the Prime Minister, Mr Menzies,
> in this case – absolute power to compel any newspaper or
> periodical to publish any statement or material supplied on
> his behalf in whatever position is required and without limit

in respect of the space occupied ... These regulations, if they were literally and arbitrarily enforced, would render the Press of Australia completely subservient to the will of the Government and the Director-General.

In his second stint as prime minister, Menzies was alive to the potential problems of free-ranging media under Cold War conditions. He took a central role in the implementation of the D-notice system and attended the first meeting of the committee that guided its development. He wrote personally to editors of newspapers and heads of media associations requesting their co-operation before the system was implemented. For example, in November 1950 he wrote to Eric Kennedy, the president of the Australian Newspaper Proprietors' Association (and influential owner of the *Sun* newspaper chain), to ascertain how an attempt to introduce D-notices would be received. He wrote in similar terms to the president of the Australian Newspapers Council, the general manager of the ABC and the president of the Australian Commercial Broadcasting Stations. In his letter to Kennedy, Menzies confirmed that the UK had suggested the D-notice system for Australia, and gave an overview: 'For many years there has been an understanding in the United Kingdom between the press and publishers on one hand, and the Defence Ministries on the other, whereby the former agree not to print, without prior reference, any matter relating to subjects specified in "D" notices'.

In Kennedy's absence, the acting president R Doutreband replied. Doutreband said he had gauged the attitude of members, which he summed up thus:

All members are anxious to co-operate with your Government to the fullest extent in ensuring the security of secret information. Their only concern has always been that measures designed to safeguard the genuine interests of national security should be used to cover political matters that have no relation to

national security. Members, however, feel that the proposals
you now put forward offer the prospect of a real and lasting
co-operation between the Service Departments and the Press
in these important matters and they will be glad to assist in the
introduction and smooth working of the scheme.

Doutreband's comments captured the mood of media proprietors at the time, all of whom seemed remarkably unfazed by the imposition of information controls. There was no such equanimity in the US, where the media actively and openly opposed any governmental controls on reporting its nuclear test program. Doutreband was rather more effusive than some, though. President of the Australian Newspapers Council Frank Packer replied rather more briefly:

Dear Mr. Prime Minister,

Further to your letter of November 22, regarding Security of
Defence Information, the members of my Council are agreeable
to this proposed Committee.

The new Australian system given the go-ahead by Australia's media proprietors was to be managed by a committee co-ordinated by a civilian from the Department of Defence – AE Buchanan was eventually appointed to this role – and chaired by the minister for Defence Philip McBride. Menzies insisted that the existence of the committee and its D-notices be kept secret (as it was until 1967) and requested that only senior representatives of media organisations be invited to join the committee. The plan was to give all the leading media organisations of the day a seat at the table, including representatives of the principal daily and periodical press associations, news services, the ABC and commercial broadcasting interests that ran independent news services. The organisations that provided the bulk of public information to the average Australian quickly and readily agreed to censor what they published.

A secret cable from Frederick Shedden to Major General Rudolph Bierwirth, the Australian Defence Department's representative in London, reported the positive response of media proprietors. Shedden said that 'consideration is at present being given to the constitution of machinery for introduction and operation of scheme' and requested that Bierwirth obtain copies of current UK D–notices to use as a guide for formulating Australia's own notices. He suggested that Rear Admiral (retired) GP Thompson, secretary of the UK committee charged with overseeing D–notices, would be able to assist. The Australian system was to be modelled closely on the UK system, and the two committees maintained close co-ordination for many years.

The Australian Government seemed prepared initially for the media to be less willing. A Department of Defence file containing information about the process to establish D–notices contains documents relating to the US that reveal some fundamental differences in approach. A resolution adopted unanimously by senior US media representatives in the late 1940s, as Cold War tensions increased and the US stepped up its atomic weapons test program, expressed a commitment to unhindered coverage of matters of national importance, while recognising a responsibility not to give away secrets that might harm the national interest.

> Conditions in the world today require the perfection of our national defense, an important part of which lies in the fields of scientific research and development of new military weapons. Protection of necessary military secrecy in such fields in a country rightfully jealous of its free and uncontrolled media of communications presents a problem in national security. We recognize the existence of such a problem. Its wise solution is the responsibility of the National Military Establishment. But it is shared to a degree by all media of public information. As representatives of such media we have willingly assumed our proper part of that responsibility. *We do not believe that any kind*

of censorship in peacetime is workable or desirable in the public interest.
If any exists, we would not be sympathetic with an intent, on
the part of the Military Establishment, to propose peacetime
censorship [emphasis added].

While explicitly rejecting peacetime censorship, the US media representatives recommended regular consultation about national security issues between the media and the US security services through
a Security Advisory Council. However, the Melbourne *Herald*
reported on 8 March 1948 that there were some objections in the
US to a regular security consultation: 'It is complained that it would
open the door to Government censorship in peacetime'.

Nevertheless, the heads of the major American press and broadcasting associations, along with representatives of specialist media
sectors such as Perry Githens, editor of *Popular Science Monthly*, agreed
to the resolution and, in doing so, specifically rejected censorship.
The first secretary of Defense James Forrestal issued a statement to
coincide with this declaration, endorsing a policy that would ensure
'full release of all possible information to the American people',
while at the same time protecting 'information which should not
be revealed to potential enemies'. Secretary Forrestal also 'expressed
himself as in accord with the declaration against censorship, since it
coincided with his own views'.

A D-notice system in the US would have been unthinkable.
The apparent eagerness of senior Australian media representatives
to join the D-notice committee suggests a distinctly different media
culture in Australia. Media freedom in the US, of course, accorded
with its constitutional guarantees of free speech – guarantees that
do not exist in Australia. And while declarations may not necessarily
reflect what happens in reality, it does seem that the American media
enjoyed greater (though not total) access to nuclear test information
than their Australian counterparts. They were also more assertive in
pursuing their rights to this information and less hindered by official mechanisms such as D-notices.

Despite across-the-board approval, the process to adopt a D–notice system in Australia took nearly two years. The delays were caused in part by the difficulty in finding a suitable secretary for the committee, disagreements as to whether the proposed system would apply to foreign media agencies operating in Australia and the exact wording of the first atomic tests D–notice.

The mechanism for the release of official information in the first D–notice, for the Hurricane test in October 1952 at Monte Bello, was initially so confused it even drew in the prime minister, who showed direct interest in the smooth working of the embryonic system. The general manager of John Fairfax & Sons, publisher of the *Sydney Morning Herald* and *The Age*, among others, wrote to Menzies in August 1952 to complain that the navy's public relations team in Sydney had rebuffed his reporters when they had attempted to obtain officially approved information about the first atomic test. The system fell apart because the contact person, Lieutenant Commander Dollard, was on leave. The general manager wrote that they would 'give the utmost co-operation to the Government in preserving security on all secret information about the tests, but it is essential that the fullest facilities should be available for us to consult responsible officers at all hours'. In a handwritten comment at the bottom his copy, Menzies wrote, 'Ask Mr. McBride to enquire into this urgently. It sounds very bad'.

As Defence minister, McBride had primary ministerial responsibility for establishing the D–notice system (with a surprisingly large amount of interest and input from Menzies), and he advised his Cabinet colleagues on the proposed system in a briefing letter dated July 1952. He listed the members of the committee as the permanent heads, or their senior deputies, of Commonwealth departments of Defence, Navy, Army, Air, Defence Production and Supply, as well as senior executive officers of the press and broadcasting associations.

The first meeting of the committee, to be known as the Defence, Press and Broadcasting Committee, was held on 14 July 1952 at Victoria Barracks in Melbourne. Before the meeting, the Joint

Intelligence Committee (the directors of security for all branches of the Australian armed services) had devised draft D-notices for the new committee to consider. These were presented as one of the main orders of business at the meeting. McBride chaired the meeting, and Menzies, who also attended, addressed it to express 'the Government's appreciation of their willingness to co-operate with the Defence Authorities in the introduction and operation of a system of "D" Notices'.

In addition to the initial set of D-notices, McBride had advised in his briefing that the first meeting would 'consider the principles upon which the organisation is based' and 'the procedure for its operation'. The seven principles adopted by the committee in due course emphasised the need to prevent dissemination of information 'detrimental to national security' and the voluntary nature of the notices, among other things. This voluntary principle has been affirmed throughout the life of the D-notice system in Australia. The most recent documentation from the D-notice committee, dating from its last meeting in 1982, stated, 'The system is an entirely voluntary one, offering advice and guidance only. Non-observance of a request contained in a Notice carries no penalties. In the end, it is for an editor to decide whether to publish an item of information, having regard to national security requirements'.

The principles noted that the media representatives on the committee would have access to information 'of a secret nature', even though only information classified 'confidential' would appear in the official notices. This trust of participating media representatives appears to have been justified. Throughout the British tests there were no known Australian breaches of the atomic test D-notices.

A major agenda item at the inaugural meeting was Operation Hurricane, the first nuclear test in Western Australia. McBride explained that it was primarily a British operation and that Australia would play a secondary role. But while the UK Government would lay down security conditions, there were some issues around media liaison that needed clarifying, and 'the Australian Government was consulting the United Kingdom Government in regard to the release of certain background information regarding activity in Australia in respect of which the Australian press had a special interest'. The Australian media were endlessly frustrated with the fact that the British media seemed to have privileged access to information about the British nuclear test program.

At the first meeting, this problem, which was already apparent and would become more so, came to the surface: British media were getting hold of information ahead of Australian media. The Australian government officials undertook to ensure that any release of information was simultaneous. (Issues over the perceived preferential treatment given to UK media festered throughout the British atomic test program in Australia and were never really resolved.) The media also asked to be present as observers, something of which Defence Minister McBride 'took note', while maintaining that under the present arrangements even he was not allowed to be present. The committee agreed on eight D-notices to be issued. These covered:

» UK atomic tests in Australia;
» aspects of naval shipbuilding;
» official ciphering;
» the number and deployment of Centurion tanks;
» troop movements in the Korean War;
» weapons and equipment information not officially released;
» aspects of air defence; and
» certain aerial photographs.

The D-notice applying to the Hurricane test at Monte Bello, originally D-notice No. 1, was ultimately designated D-notice No. 8 after being altered to take into account media objections both at the meeting and in subsequent correspondence. Shedden reported just after the meeting, 'Everything went well except that, as expected, the arrangements for publicity in connection with the atomic test were

215

somewhat critically discussed ... [The media at the meeting were informed] that we were seeking to liberalise the United Kingdom outlook in so far as treatment of the Australian press is concerned'.

The hard line that the UK authorities took on co-operating with the media applied particularly to the Australian media. The British media – arguably because of more sophisticated reporting skills – displayed a greater depth of coverage in their stories of the tests and regularly 'scooped' Australian reporters, a fact that caused problems for the minister for Supply Howard Beale, who faced complaints from Australian journalists.

Following resolution of the criticisms aired at the first meeting, the revised and re-numbered D-notice No. 8 was officially distributed to the media just before the Hurricane test. In it Captain AE Buchanan, secretary and executive officer of the committee, set out a number of restrictions in reporting the Monte Bello test. These included that there be 'no disclosure of, or speculation concerning' information regarding the technical details of the weapon design, the precise form and date of the trials, the results to be obtained and the passage arrangements for the fissile material and for 'the Main Force after leaving the United Kingdom, until released by the United Kingdom authorities'.

These were sweeping restrictions. The notice also spelled out what information would be made available 'unofficially' to the media by the navy's public relations office, including such items as the transfer of a construction squadron of the RAAF to Monte Bello and the build-up of stores at Fremantle to the south of the test site. The statement exempted from the notice matters of 'observable facts which must inevitably be known to foreign observers and the press'. These included arrivals and departures from ports and airfields but excluded anything taking place within the 'prohibited area' of the test itself.

The notice, in effect, limited the media to what they could directly observe by stationing themselves close to the test site, and to what the test authorities chose to tell them. It specifically, and not surprisingly, precluded any technical detail about the atomic device itself. Obviously, it did not restrict political speculations about the wider meaning of the bomb test and the development on Australian soil of a British nuclear deterrent, although few such broad stories appeared in the media. The primary purpose was to restrict the promulgation of technical and strategic information.

Some Australian media organisations quibbled over the delivery of information, appeals that did not go completely unheeded in Canberra. Menzies wrote to all media organisations in August 1952 to assure them that their views on media coverage of Hurricane were being taken into account. 'As the test is of great public interest the Government expressed to the United Kingdom Government its view that any possible information that could be given to the press without prejudice to security, should be made available.' The Defence, Press and Broadcasting Committee had discussed this issue at its first meeting the month before.

Menzies also claimed that the problem of the UK media gaining access to bomb test stories before the Australian media had been solved through his government's representations. The Australian press should now receive the same treatment as the UK press regarding 'the release of official information and photographs'. (Despite Menzies' intervention, though, the Australian media continued to complain about perceived preferential treatment for the British media.) He attached the revised D-notice, which Buchanan had issued a little earlier. The first, disputed version had omitted the list of 'background information' items included in the final version, which gave media access to substantially more information than originally offered. The changes were reasonably well received, although the notice still had a sticking point, namely the line that restricted reporting on 'anything on or taking place in the prohibited area'. As the editor of the *West Australian* newspaper EC de Burgh pointed out:

How are we to know that something may not be clearly audible
and visible from the mainland and be seen and heard by scores
or hundreds of observers who may include representatives of
foreign, even Russian, newsagencies? Why should we be asked
to agree not to publish tomorrow something already known
to, possibly, hundreds of West Australian civilians and may-be to
dozens of foreign or Communist observers?

Despite de Burgh's heated letter, the wording of D-notice No. 8 was
not further changed before it was issued in the lead-up to Operation Hurricane. On 10 November 1952, after Hurricane was over,
Buchanan cancelled this first atomic-related D-notice and thanked
the media for their co-operation, 'which was an important contribution towards safeguarding defence information in connection
with these tests'.

While the Australian media generally accepted the D-notice
system without qualms, there were some hiccoughs. Press agencies such as United Australian Press and Australian Associated Press,
which gathered and sold stories to a wide variety of outlets including those overseas, refused to co-operate with the new atomic
weapons D-notice, leading to a prickly relationship. In a minute
paper sent to Shedden, committee secretary Buchanan wrote that
in response to his direct inquiry whether they would co-operate in
D-notice No. 8, Mr Richards from United Australian Press said 'no',
while Mr Hooper from Australian Associated Press 'gave an equivo-
cal answer, that was in fact a negative'.

The agencies sold stories to outlets overseas that were not
bound by the Australian D-notice system. If they limited their coverage, visiting and unfettered foreign journalists could trump them.
Of the eight D-notices agreed by the committee, the British tests
were of greatest interest outside Australia. The agencies would participate only in D-notices that applied to other defence matters
and not to the big, newsworthy story that most interested them.
To get around this, Buchanan suggested, and Shedden agreed, that

the main news agencies 'including American agencies but omitting other foreigners' be asked to participate in the D-notice system after Operation Hurricane had concluded. The position of the Australian agencies remained ambiguous, and their representatives did not join the D-notice committee in 1952 or subsequently. Also, negotiations with the American agencies broke down, and they were never included in the D-notice system.

The British had had the same problems with the US press agencies in relation to their own D-notice system. Even so, in practice the British D-notice committee had found the US agencies tended to abide by D-notices sent to them informally, while making an outward show of their independence. However, George Thomson pointed out some exceptions, including two US air journals with a fairly large sale in Britain which were constantly publishing information about British military aircraft that British editors and air correspondents knew perfectly well but did not publish because of the D-notice. A well-known Swiss air journal was an even worse offender. 'All this caused much complaint and bad feeling among British editors.' Thomson recounted a remark made to him by the head of the United Press of America, who acknowledged that the US press gave the Russians 'a great deal of confidential information about the U.S. armed forces' but said he would still 'fight anybody who attempted to restrict the freedom of the press in any way'. Thomson added two exclamation marks at the end of this quote.

The establishment of D-notices was a logical step for the Menzies government, to please Britain and to keep the Australian media under a measure of control. However, publicity also had propaganda benefits. Exactly where to draw the boundary between public and secret information was a difficult task that exercised Menzies and his colleagues – especially difficult since they had to adhere to the wishes of the UK authorities. Media information was often initiated in Britain for later distribution in Australia, and the British D-notice committee issued its own notice for the atomic tests, applicable to British journalists. The British and Australian committees consulted closely and synchronised their activities in relation to the nuclear tests (and possibly other things too).

Sometimes these mechanisms created frustration for all media, particularly overseas representatives. Indeed, for officials from the Australian Department of Supply (which managed the Australian Government public relations for the tests) relations were frequently more strained with the overseas press than with domestic media, and foreign media representatives were not above a bit of manipulation to get access to information. For example, in a letter to an Australian Government public relations official, the chief correspondent and South Pacific manager for the US agency United Press Associations GE McCadden objected to plans for strict exclusion of media from the Hurricane test: 'If such a policy is pursued, it is my belief that the U.S. Press is most unlikely to devote as much space to the tests as it would utilize were your authorities to relax these announced restrictions'. McCadden went on to point out the benefits of gaining well-informed American media coverage, one suspects in much the same way that American media made their case domestically:

> If one of your ultimate major objectives of these tests is
> to impress upon American public opinion a spectacular
> achievement of our major ally which contributes to our
> common strength, then the best means of reaching such public
> opinion is through the eyes and ears of American reporters,
> including United Press.

McCadden also said that the 1200 newspapers, 1100 radio stations and 50 television stations served by his agency would all take the United Press Association's news stories about the tests 'regardless of how [the agency] gets the information about them'.

This forceful, eloquent and at times impassioned case for media access to Hurricane comments on the generous access granted to British and Australian correspondents to the US atomic tests at Bikini Atoll in the Pacific. He concluded, 'I want to stress … that all I can tell you now is how the strategic U.S. Press will react to a continued news blackout on one hand, and a new policy of relaxing that blackout on the other hand'. The thrust of this letter was unmistakable: the potentially positive message of the atomic tests could be distorted if the Australian authorities did not provide the media with what they wanted. In the 1950s American reporting of their own atomic tests was more thorough and critical than any coverage that appeared in Australia.

In 1953, after a period of negotiation, the Australian Government had to admit defeat. Frank O'Connor, secretary of the Department of Supply, said in a letter to Major General EL Sheehan, Australian defence representative, that because of the unfavourable attitude of the US press agencies, the D–notice 'Committee's recommendation, with which the Minister concurred, was that the United States Press Agencies should not be included in the "D" Notice system at this stage'. D–notices were to be confined to UK and Australian media. The Americans were exempt.

Buchanan issued a new D–notice for the Totem tests at Emu Field in October 1953. It had a similar structure to the Hurricane notice but added a more specific restriction, on 'nuclear efficiency and measurements relating to weapon efficiency', further limiting the dissemination of technical detail as the design of British nuclear weapons became more sophisticated. This D–notice also listed a number of new things that would be provided to reporters as background material, including an initial survey of the area by Sir William Penney, the work of construction personnel, assistance given by the LRWE at Woomera, air–lift operations by Yorks and Bristols, boring operations for water and study of geology, the work of Australian scientists in checking margins of safety, the transport of aircraft and war stores to the site and the co–operation of pastoral lessees.

On 26 June 1953, just before the Totem D–notice and when Menzies was overseas, the acting prime minister Arthur Fadden distributed a letter to the press representatives on the Defence, Press and Broadcasting Committee. This letter expanded upon their need to restrict information about the new mainland test series and explained that journalists would not be allowed to witness the Emu Field tests. Fadden made the case that press exclusion from the test site was intended to eliminate media pressure to hold the test before the conditions were right:

> It is desirable that the man in charge of the operation should have a considerable margin of time to play with as to when the test should take place. The presence of the press has a tendency to lead to attempts to meet a scheduled date and this could cause a reduction in the value to be derived from the test.

While delays in testing, particularly during Totem and later Buffalo, did make the media restive, the media did not influence the exact timing of any tests, although they probably exerted indirect pressure. Howard Beale was asked a question in parliament on this issue at the height of media unrest at the time of Buffalo in 1956 and replied, 'I can say that there will be no change in the standards of safety which have been, and will be, maintained from first to last in conducting the tests'.

Fadden's argument did not convince the media chiefs, who lobbied to gain access to the test site to view Totem, and on 10 September permission was granted. The D-notice system had already done its job by drawing the media into the preparations for test publicity. This concession to media demands suggests a powerful co-operative process that seems to have been working the way it was intended.

After Totem, the operation of D-notices becomes less clear. In 1955, Frank O'Connor sought clarification on the arrangements for a D-notice for Operation Buffalo in September 1956. He wrote to his counterpart at the Prime Minister's Department, Allen Brown, to find out what was happening. 'It is highly desirable that any such "D" notice contemplated for "Buffalo" or any other similar operation should be accepted by the Committee well in advance of any public announcements of the trials, and before activity has reached a level which draws attention and subsequent press comment.'

A variety of almost indecipherable squiggles in different hands festoons the lower portion of this letter, including one stating, 'This matter can't be dealt with until ...' followed by a tantalising but unreadable reference to the 1956 Monte Bello tests, Operation Mosaic. This frustrating note is tagged with what looks like the date '29/8/55'. The final handwritten comment states, 'Matter Completed 16/9/55', which was a year before the first Buffalo shot at Maralinga. The official National Archives D-notice files contain no further documentation on D-notices for the British nuclear test series, and it seems likely that the notices were specifically issued only for the earlier tests (definitely Hurricane and Totem, and possibly Mosaic). The earlier D-notices may have been deemed to be current for the later tests. Arthur Fadden in his letter to press chiefs indicated that the Hurricane D-notice also applied to Totem, despite the fact that it had already been explicitly cancelled.

Once the Defence, Press and Broadcasting Committee decided on its early D-notices, it rarely met again. The committee operated independently of other Australian security authorities, so its concerns were mostly those of its parent organisation, the Department of Defence, not those of the other government entities concerned with security. The British D-notice committee regularly consulted with MI5, but Australia's D-notice committee never seems to have had a similar relationship with ASIO. Indeed, the Joint Intelligence Committee reported that ASIO rebuffed its suggestion that it participate in the formulation of the initial draft D-notices: 'The Australian Security Intelligence Organisation did not propose to submit draft "D" Notices at this stage because only in the most exceptional circumstances would the need arise for that organisation to sponsor a "D" Notice'.

The D-notice system remained secret until Prime Minister Harold Holt confirmed its existence in October 1967 in response to a media article by the journalist and lobbyist Richard Farmer. In

November, Holt answered a series of questions in parliament about D-notices. In fact, Holt's comments on D-notices must be among the last parliamentary questions he answered before he drowned in mysterious circumstances in December 1967. One presumes there was no connection.

The Defence, Press and Broadcasting Committee still notionally exists but has not met since 1982 and is unlikely to do so again. At the 1982 meeting, the committee considered all seven remaining D-notices, which did not include atomic tests, and reduced them to four. Those four are still, technically, in effect and refer to the capabilities of the Australian Defence Force, including aircraft, ships, weapons and other equipment; the whereabouts of Mr and Mrs Vladimir Petrov (no longer relevant as both are now dead); signals intelligence and communications security; and the Australian Secret Intelligence Service. Perhaps not surprisingly, the intelligence service submitted to the 1995 commission of inquiry into its operations that a replacement for the old D-notice system was needed. In a rather sour official submission to the inquiry, the service said:

> The current D-notice [system] is inadequate because it relies on voluntary media restraint, which no longer exists. Changes in Australian society since the 1950s have led to debate as to how principles of public perception and independence of the media can be reconciled with secrecy required for the sake of national interest. This debate has engendered increasing disagreement on what constitutes the national interest. The media organisations have shown by their actions that they will decide what the public interest is in any given situation without assistance from those affected. The media organisation's [sic] perception of the national interest appears to coincide with its own journalistic interests.

It took 15 years for the system itself to become public knowledge. By the time it did, as legal academic Laurence Maher said, 'the Australian media was becoming more probing and diversified'. D-notices have no further influence over what Australian media publish or broadcast. As an interesting aside, however, the Labor federal government proposed an updated form of D-notices in 2010, in light of rising terrorism concerns and the WikiLeaks disclosure of sensitive diplomatic and military information. The suggestion was quickly abandoned.

The D-notice system established a formal co-operative relationship that set specific reporting ground rules that, for the most part, the media seemed willing to obey. The notion of media restraint and the prerogative of government to keep certain designated facts out of the media – with the agreement of the media themselves, secured in a committee that included senior media people – affected the way the media reported the tests. This had a cumulative effect. D-notices were not issued for the most dangerous activities at Maralinga, the Vixen B experiments in the early 1960s, but, by then, the media were in the habit of reacting to government-approved media releases. It did not seem to occur to them to investigate stories on their own account.

The conditions of the time were conducive to secrecy, and government policies strengthened those conditions. The Australian Government, with backing and pressure from the British Government, used D-notices and other information controls to restrict media scrutiny. Compliant media conditioned to receiving government information in, mostly, a controlled and predictable way were reinforced by an official, but not legally binding, system that forbade reporting certain secret activities. When the Geneva moratorium on weapons testing came into effect, the British banished Maralinga from even the weak media spotlight. The ill-equipped media did not pursue the story because they did not understand its complexities and implications. The minor trials continued for several years without media attention. No wonder the British nuclear tests in Australia remained mysterious for decades.

9

Clean-ups and cover-ups

*Long-term or permanent habitation of contaminated areas is
improbable even in the distant future.*

Atomic Weapons Tests Safety Committee, 1967.

*Eames: When it came to the clean-up exercise, was the situation
this: that the Australians had absolutely no way of knowing what
the debris was that would have occurred from these tests apart
from what you told them?*
Pearce: That is so.

Geoff Eames, counsel for the Aboriginal people at the Royal
Commission, and Noah Pearce, AWRE scientist, 1985.

*There came a point, when Sir Ernest Titterton was giving evidence,
when there was almost no point asking him anything, because we
could not get the facts out of him.*

Justice James McClelland, in an aside to John Moroney
during Moroney's testimony about Operation Brumby at the
Royal Commission, 1985.

What do you do with a vast nuclear weapons test site that is now surplus to requirements? The Partial Test Ban Treaty of 1963 made official what the moratorium of 1958 to 1961 had begun – there could be no more legal atmospheric tests of nuclear weapons by nations who signed the agreement. Maralinga was built for atmospheric tests. Between the end of the moratorium in 1961 and the start of the treaty in 1963, there was even some thought that Maralinga might resume a major trials program, especially if the touchy Americans pushed the British away from Nevada again. However, the treaty came in and Britain worked with America in Nevada and the Pacific. Maralinga was history.

The British did canvass the possibility of taking the tests underground and hinted at a site about 400 kilometres from Maralinga where a hill rose 700 metres above the plain, generally considered to be Mt Lindsay. By now, though, the Australians had had enough. The hill looked like it was in an Aboriginal reserve, and drilling into it would, at the very least, likely cause water flow problems in the area. The final meeting on the subject took place in December 1963 between senior members of the Prime Minister's Department and their counterparts from the UK weapons establishment and the UK High Commission. The talks came to nothing. The Australian Government was a signatory to the Partial Test Ban Treaty and in no mood to test its limits, or to contend with a restive populace who no longer thought atomic weapons testing desirable. The idea of underground tests quietly died. The Menzies government was relieved.

No-one really knew what to do about Maralinga though. The British kept their options open for a while and took several years to close it down fully. The original agreement to allow the British to test atomic weapons at Maralinga committed them to clean up what they left behind. Clause 12 made the British liable 'for such corrective measures as may be practicable in the event of radio-active contamination resulting from tests on the site'.

The first attempt to put the site right was called simply Operation Clean-Up, a basic name for a basic operation organised by the range commander. Just about all remaining personnel on site pitched in. Every Tuesday afternoon from 25 June 1963 for a number of weeks, they carried out an 'emu parade', picking up scraps of various kinds. Site personnel gathered 175 tonnes of contaminated material from the three Naya sites, as well as TM100, TM101 and Wewak. This debris was all placed into a pit at the 'cemetery' at TM101. (As a side note, there was no actual cemetery there, or at the airfield. The name was an in-joke about a burial place for inconvenient items.)

Also in 1963, AWTSC secretary John Moroney reviewed radio-active contamination at the site. He reported on 5 September 1963 that the plutonium used in the minor trials was the most danger-ous hazard, particularly that at Taranaki left over from Vixen B. He also found quantities of the bone-seeking radioactive element strontium-90 at the site. However, he was hamstrung by not having detailed information from the British, particularly data to do with the minor trials. The British had shared virtually no information about the minor trials with the Australians, even when the Department of Defence reacted against the Titterton-inspired information bottle-neck. Moroney wrote to the AWRE in November 1963 seeking more information, but, at that stage anyway, it was not forthcoming.

The biggest problem left behind was plutonium. That word alone should have been enough to ensure a thorough clean-up. Make no mistake; the dangers of plutonium were well known in the 1960s – ignorance does not explain the persistence of the contamination problems at Maralinga. These might not have been so severe, either, if the British had not whitewashed the true state of the range via a nondescript document called the Pearce Report.

Noah Pearce, the author of the report, looms large in this part of the story. Pearce had an honours degree in physics and had worked during the latter years of World War II on measuring the effects of bomb blasts for the UK Ministry of Supply. He lived a long life, dying in 2009 at the age of 91. He worked for the AWRE in various capacities, particularly in the early days measuring the explosive yield of Hurricane and Totem. In the late 1950s he was responsible for health aspects of the minor trials and, later, for the clean-up operations: Operation Hercules in 1964 and Operation Brumby in 1967. Both these operations made the contamination problem worse, and both seem to have been conducted largely for the sake of appearances, rather than to actually clean up the appalling mess at the site. 'Operation Brumby', said the Royal Commission, 'was based on wrong assumptions. It was planned in haste to meet political deadlines and, in some cases, the tasks undertaken made the ultimate clean-up of the Range more difficult'. Later, a report by the Maralinga Rehabilitation Technical Advisory Committee (MARTAC) on the rehabilitation of the Maralinga site stated that Hercules and Brumby 'did not rehabilitate the site to the standard later recognised to be necessary for the protection of people and the environment'. How could something so important be so wrong?

For a number of years, the Pearce Report was held to be the final word on the radioactivity at both Maralinga and Emu. Never has there been a greater example of the bureaucratic convenience of an official report – a report that's completed, checked, signed off, printed, filed and then forever taken as true. Well, not actually forever in this case. The Pearce Report is a pivotal document because its contents were the backdrop to the formal agreement to hand the site back to the Australian Government. The British and Australian governments both clung to it as to flotsam from a shipwreck for way too long, even though the evidence suggests that the Australian officials had not read it closely. In fact, the Pearce Report was barely the first word. That it was *not* forever taken as true came down to a combination of whistleblowing and investigative journalism.

The first Pearce clean-up, known officially as Hercules 5, arose from the growing conviction that Maralinga would never again be used for atomic weapons testing, and therefore the range was likely to be left unattended and unmaintained indefinitely. Key on site personnel such as the Australian health physics representative Harry Turner had already been relocated (Turner was transferred to the Department of Defence in Canberra). There was no health physics expert on site, no-one to deal with a radiation emergency or keep an eye out for untoward scatterings of cobalt-60 beads.

The AWRE formed a small decontamination and health physics team in the UK and shipped them to Maralinga to work with range staff on Hercules 5, under the supervision of Pearce, who was on site for the last three weeks of the operation. There was only one Hercules operation. Pearce said the name Hercules 5 was 'dreamed up by one of the staff because the fifth labour of Hercules was to clear out the Augean stables'. In classical mythology, this labour was, tellingly, considered to be both humiliating and impossible. Counsel assisting the Royal Commission, Peter McClellan, mused aloud that perhaps the AWRE saw this part of Australia as Augean stables. Pearce kept his answer neutral and non-committal. But there was a heaviness in the atmosphere during Pearce's testimony in London. The Royal Commission was in the UK capital to question the people responsible for radioactive contamination of Australian territory. Pearce's name was inextricably linked to this dreadful aftermath.

The Hercules 5 clean-up plan was presented to the AWTSC in 1964 and passed with few amendments. The Royal Commission later dwelt on the fact that the clean-up and survey phases were rushed through, noting how the AWTSC had enthusiastically rubber-stamped the plan and no-one had stopped to think that one day Indigenous people might want to come back to the lands from which they had been so brutally expelled. Also, while approval was given for cleaning up within the boundaries of the range, no plans were drawn up for surveying and clearing up any contamination that had fallen outside the boundaries, other than at Emu.

During his grilling at the Royal Commission, Pearce constantly put the responsibility back on the AWTSC, minimising his own input in the management of the clean-ups and survey. He made it clear that he had not thought much about the plight of the Aboriginal people, and it had not informed his approach to his task. The problem was that the AWTSC did not want to manage the clean-up either. No-one wanted to own the responsibility for a dying weapons range. Both the AWRE and the AWTSC thought the Maralinga era was over and would soon be forgotten. The AWTSC explicitly stated in its liaison with Pearce, as revealed in testimony to the Royal Commission, that once the range had been tidied up it should be patrolled for about 15 to 20 years. After that time, the public would no longer be interested in it. When the Royal Commission raised this time frame with Pearce, he did not see anything particularly cynical about it, though he pointed out, a little ruefully, that interest had not in fact disappeared after all. On the contrary.

Hercules 5 was not based on a physical survey of the site but on existing records. That meant making a number of lazy assumptions rather than plotting the real radiation risks on the ground. The task set for the Hercules team had 10 parts to it, encompassing cleaning contaminated buildings and vehicles, burying any contaminated materials that could not be cleaned in pits at Taranaki, carrying out a health physics survey of the area to determine the risk to human health of any contamination, fencing any areas that still had residual contamination and preparing drawings to show where the fences and pits were located.

Hercules lasted from August to November 1964 and produced two reports, one describing the radiological state of the range at the conclusion of the clean-up and the other prepared especially for the AWTSC describing the residual radioactive and toxic contamination still present. Pearce stated at the end of the operation that the signs and fences would be regularly inspected and maintained, but that the range was now more or less in a holding pattern, pending a decision on its future. He recognised that Hercules was not a final clean-up but thought it would do until longer term decisions were made and personnel departed the site.

Pearce well knew that plutonium was loose on parts of the range and, with his team, calculated the inhalation hazards from the plutonium mixed with the dust. The inhalation hazards drove much of the clean-ups. He also recommended, based in part on data provided by Harry Turner, that the topsoil could at some point be ploughed to dilute the plutonium by mixing it through the soil to a greater depth than where it currently lay. This proposal was put to the AWTSC, which agreed that diluting the plutonium was wise. Ploughing and grading were duly carried out around Wewak, TM100 and TM101, as part of Hercules. The contents of various contaminated refuse pits around the site were consolidated into a total of 19 huge burial pits at Taranaki. The pit area was enclosed in a high chain-link fence. Ploughing was not, however, such a great idea. As radiation scientist Geoff Williams later said about ploughing the plutonium into the topsoil:

> That might have sounded very nice in the lush fields of southern
> England but out there [at Maralinga] you have three or four
> inches of sand on top of very hard limestone, so in many cases
> the scraper was just bumping along on the limestone. If it had
> solved the problem, if it had really diluted to a sufficient level,
> it would have been all right. But it didn't because the fragments
> and the concentration were so high, all it did was make a bigger
> mess.

No measurements were made of the effectiveness (or, indeed, otherwise) of ploughing the plutonium into the soil. As Pearce said in evidence, 'We knew the levels of activity on the surface and it was reasoned that if that were distributed uniformly through a thickness you would finish up with a uniformly contaminated layer of soil some inches or 15 cm thick … I cannot recall any experimental evidence for that at the moment'. Hercules was conducted without any form of testing on either side of it. It was mostly guesswork.

The original 10-year agreement for the British to use Maralinga was due to expire on 7 March 1966. This forced a decision, because maintaining the site was costly and pointless unless it was going to be used for further tests. On 16 February 1966, four weeks after Menzies retired from office, the Australian Government received word from the UK High Commission that Britain would relinquish the site. The Maralinga agreement meant that they needed to clean it up. Remarkably, the AWRE – in possession of so much knowledge about what was left there – proposed four completely inadequate actions to make the site safe, starting with disc-harrowing some open areas of ground 'with a view to reducing the hazard to a level safe for permanent human habitation'. The other three measures were sealing the pits with concrete and sand, removing from Maralinga village a small amount of radioactive ducting material and sealing drains. Disc-harrowing – churning up the topsoil using a tractor towing large circular blades – could never make the area safe for people to live in. At the Royal Commission many years later, John Moroney gave evidence that even the notoriously lax AWTSC never took this suggestion seriously. He said that the only way that the site could be made habitable was by removing all the plutonium, not by churning it into the sandy topsoil. At this point, Jim McClelland berated Moroney, and the safety committee, for not just recommending that all the plutonium be removed, instead of mucking around with ploughing and disc-harrowing. Moroney replied, ruefully, 'It would have been a good idea, yes. We would not have had this Royal Commission'.

Pearce did not spend a lot of time physically at Maralinga, although he visited at crucial moments. In early 1966 he arrived at the test range with his AWRE boss Roy Pilgrim to survey the site. At that time they decided to set up RADSUR to survey Emu Field and Maralinga so they could produce the appropriate documentation needed to meet the requirements of British withdrawal. They made no such plans for Monte Bello, which was essentially ignored until much later. RADSUR was a flawed attempt to map radioactive contamination before Operation Brumby.

Operation Brumby, the next phase of the withdrawal from Maralinga, was the main scheduled clean-up. Unlike Hercules, which had been undertaken using existing data, Brumby needed a more thorough understanding of the radiological properties of both Emu and Maralinga. The AWTSC asked the Australian Government for some direction on how to deal with the derelict range, but it never received a response. At its 133rd meeting on 14 May 1966, the AWTSC accepted the AWRE's RADSUR proposal. The AWTSC still had no idea what the Australian Government wanted, but the plan proceeded anyway. On 21 December 1966 Moroney wrote rather forlornly to the Department of Supply saying, 'In the absence of such a decision, the AWTSC will base its decision on complete evacuation of the range'. No-one in authority seemed to be interested in tying up the loose ends. Maralinga was a problem that just about everyone was trying to wish away.

RADSUR was carried out between October and December 1966 by about a dozen AWRE scientific personnel, assisted by six sappers from the Royal Engineers. The survey involved gamma and beta surveys at each of the major trial sites and the plutonium-contaminated areas around Taranaki, TM100 and TM101. With the exception of Marcoo, where the low-yield ground shot formed a crater, RADSUR noted that the major trial sites at both Emu and Maralinga had glass-like glazing produced because the intense heat from the blast had fused the soil. This strange shiny glazing covered a circular area with a radius of about 180 metres at each major site. The glass was alive with beta radiation, largely from the strontium-90 trapped in it. However, there was not much in the way of gamma radiation, and what little there was fell away sharply as one moved away from ground zero. Most of the radioactivity associated with the big bomb sites was in the upper layer of soil, between 30 and 40 centimetres deep, close to ground zero and approached the surface as the distance from ground zero increased.

RADSUR established a colour-code system that helped to map contamination. The Pearce Report used this system to explain the state of each part of the site. Yellow areas had the highest contamination – above 400 kilobecquerels (10 microcuries) per square metre – while red areas had between 40 and 400 kilobecquerels per square metre. Everything else was white. During RADSUR, the Taranaki site was sampled, as it was known to contain the most contamination. The Pearce Report later outlined this part of the operation:

> The samples were in the form of a core approximately 16 cm deep by 7 cm diameter obtained by driving a tube into the ground. The sample was divided into layers 4 cm deep and the plutonium content determined by the scintillation counting method which was used for the measurement of surface soil samples. At some sampling points the rock substratum was near to the surface of this ground and prevented the sampling tool from being driven in to its full depth. This resulted in fewer soil samples in the lower layers.

Some of the samples were taken to Aldermaston for analysis, while others were analysed on site. Ernest Titterton had travelled to Aldermaston in June 1966 to confer with Pearce, and between them they had apparently agreed to keep the sampling to a minimum. That Pearce and the AWRE wanted this done as quickly and efficiently as possible is clear from the correspondence between Pearce and the AWTSC in the lead-up to RADSUR. Moroney had been offering his suggestions about what needed to be done. In a letter to Moroney in August 1966, Pearce said that he agreed with Titterton, who, he felt, wanted to keep sampling to the minimum necessary. Pearce also told Moroney he wanted to minimise the personnel involved and suggested that CSIRO not take part, although it had offered to help.

Pearce's letter to Moroney prompted the latter to write an annoyed letter to Titterton, reacting to the suggestion that he was being over-zealous. Moroney said:

> From Noah's letter and from your one comment to me when you came back, I get the impression that both of you are somewhat fearful that I am wanting to grow the sampling into a major and disproportionate effort. Noah recalls his conversation

with you at Aldermaston and writes of 'not letting the soil
sampling get out of hand and tailoring it to the minimum
necessary to establish the conditions obtaining at the various
sites'. This is a pretty strong comment, and I certainly didn't
regard the suggestions I have made as being extravagant and
warranting such a reaction.

Titterton replied more or less endorsing a quick sampling regime,
while blaming Pearce for any misunderstanding:

> Noah's interpretation of his discussion with me in June in
> England has been translated in a very free fashion. I certainly am
> not enthusiastic to have groups of CSIRO, AAEC [Australian
> Atomic Energy Commission] or anyone else charging around
> the site. Nor am I enthusiastic for you, plus supporting staff, to
> spend a lot of time on the job ... As we discussed in the past, our
> policy should be to get sufficient information for our purposes
> but not to make a big project of it.

Moroney was not as dismissive of these issues as his boss. In a
letter to Bill Gibbs, director of the Bureau of Meteorology, he said:

> There may be several points of principle wrapped up in all of
> this. The first is that we must be sure that we are getting all
> of the information we believe to be necessary for taking the
> decisions and to provide the essential data for the archives in
> the future. The second point is really a question of whether
> we should become directly involved ourselves in the clean-up
> operations; I can see that if one of us is actually there doing some
> work whilst the AWRE team is also on site, we will obtain direct
> confirmation of what we already believe, namely, that they will
> carry out the operation very thoroughly indeed. However, it may
> be worthwhile doing this simply because of the finality of the
> whole procedure.

Was there a hint of irony in this comment? There may have been, although it was a couple of decades before Moroney accepted that the operation was not carried out 'very thoroughly indeed'.

Apart from sampling, other forms of measurement were undertaken. For example, RADSUR deployed an x-ray device. The monitor had a strap that the user slung over his shoulder and a jig to standardise the distance it was held above the ground. The user would stand still every 90 metres and wait for the needle on the meter of the instrument to arrive at a steady value. Once it stopped, this number would be recorded as counts per second. The grid was coarse since it would take too long, even with this over-shoulder device, to sample more frequently. Later events proved this was unfortunate since it missed much contamination.

For all its up-to-date measuring equipment, RADSUR did not capture the right information because its field data gathering was not properly thought through. Moroney came to the view that Operation Brumby was based on RADSUR data that were 'so poor as to be useless'. Nevertheless, RADSUR formed the basis of Operation Brumby, which was intended to put the atomic test sites to rights.

Emu Field also underwent a cursory clean-up as part of Operation Brumby. The Totem bombs at Emu had produced local fallout, and, in addition, the Kittens initiator tests had left behind beryllium and polonium. The grey metal beryllium is chemically toxic, though not radioactive. It should have been removed from Kittens test sites at both Emu and Maralinga. It was not. By the time of the clean-up, the polonium, which is radioactive, had decayed. When the ARL team visited in 1984, radiation levels from both Totem and Kittens had dropped, but Emu was still unsuited for continuous occupancy and will be until about 2025.

The Pearce Report documented, quite briefly, the actions taken to survey and remediate the many sites of bomb tests. The 51-page report had numerous diagrams showing schematically what was done at each site. At Emu, the Totem 1 and 2 sites were hand-scavenged to remove metal debris and the larger pieces of glazing. An area of approximately 130 metres radius was graded and disc-ploughed at each site. All of the fences and signs put up during Hercules 5 were removed in an effort to return the site to its pre-test appearance – to spirit away the British presence.

At the various Maralinga sites, much the same process was undertaken. Most of the fences and warning signs were removed, debris and glazing were removed by hand, and there was much ploughing and grading. In addition, soil was brought in to cover ground zero to a depth of a few centimetres. At Marcoo, where the ground-detonated bomb had made a crater, about 1.5 metres of earth was dumped and levelled, to even it out. Two new waste burial pits were dug into the limestone, making a total of 21. All the burial pits at Taranaki were capped with reinforced concrete.

The minor trial sites at TM100, TM101, Wewak and Taranaki were either yellow or red areas. They were ploughed and covered with topsoil. The fences and signs around the Taranaki pits were left and the concrete covering the pits had 'RADIO ACTIVE MATERIAL BURIED HERE' imprinted onto it. Some of the contaminated soil from Wewak found its way to the Marcoo crater as fill. Several pits at TM101 were capped with concrete, like the Taranaki pits, and fences and warning signs were erected.

The Pearce Report made clear assertions about the safety of the site:

> At One Tree, which has the highest doserate, a member of the public could, in 1967, stand up to four days a year continuously at ground zero without exceeding the dose limit recommended by ICRP [International Commission on Radiological Protection] …
> At the present time, even a permanent inhabitant of the Range, free to move about at will, would not be exposed to a significant radiation hazard unless he chose to spend most of his time at or near a ground zero. This eventuality is most improbable now and the likelihood of its occurring in the future should be considered in light of the above comments on the decaying gamma doserates.

Pearce identified Taranaki, TM100 and TM101 as the worst parts of the site. 'Following the treatment given during Operation BRUMBY the only sites where the level of radioactivity warrants further consideration are Taranaki and TM100–101. At all the other sites the contamination is well below the recommended permissible levels.' Later, Pearce admitted to the Royal Commission, 'I concede that it would have been useful, with hindsight, to have examined those very high readings more closely'. Instead of increasing the measurements in the vicinity of the high-plutonium areas, Pearce and his team simply stuck to the same system used across almost the entire range, taking a measurement every 90 metres (although, anomalously, TM101 was surveyed every 9 metres). As Geoff Williams and the other members of the ARL team discovered in 1984, an awful lot of plutonium slipped through that large-gauge net.

The Pearce Report claimed that about 20 kilograms of the 22 kilograms of plutonium had been buried, thus rendering it less dangerous. In fact, an estimated 20 kilograms was later found scattered around the site, in the form of particles of various sizes dispersed locally and fine dust spread over a very large area of outback Australia. Moroney believed that the inadequacies and gross inconsistencies of the plutonium field data used for Operation Brumby were not simple mistakes. He later discovered, as discussed in chapter 11, that the AWRE knew of the errors involved. These errors resulted in considerable confusion and misinformation about plutonium contamination at Maralinga for many years, despite Moroney's vigorous efforts immediately at the conclusion of Vixen B.

Pearce never publicly admitted any flaws, and his Royal Commission statement claimed, 'I would say that Operation Brumby was a successful exercise, meeting in all respects the detailed requirements of the [AWTSC] who were kept fully informed throughout the preparation and implementation of the proposals, and who were satisfied with the results at the time'. His statements were challenged at the Royal Commission when it sat in London in March 1985, and Pearce spent two uncomfortable days being questioned by Counsel

Assisting Peter McClellan. His frequent tussles with McClellan, and his protests that he could not remember any details, added to a sense that his was not a firm hand on the clean-up tiller.

The tone of the interaction between McClellan and Pearce was often testy and strained. Pearce was taxed on the tricky convolutions that the AWRE nuclear elite engaged in to slip Vixen B under the radar of the nuclear weapons test moratorium and was also asked about the ineffectual clean-ups. He was questioned repeatedly about whether he or others involved gave a thought to the way the site might be used by people in the future, and whether cost drove decision-making. He said, 'My understanding is that at the end of Brumby we had satisfied the requirements of the Safety Committee and could declare that the range had been controlled'. He stuck to this line throughout his evidence to the Royal Commission, even though the story had already started unravelling. To the end, even with the evidence piling up in front of him, Pearce maintained that he had left the site in a safe condition and that anyone visiting the site would not be harmed. Would he allow his children to eat a rabbit caught on the range? He would eat one himself, he replied.

In March 1967, George Owen arrives at Maralinga straight from his posting in West Germany. He will turn 26 the following month, a long way from home. He is a plant operator in the British Royal Engineers and able to manoeuvre heavy machinery. He volunteered to come to Maralinga because he was bored and fed up with his service in Cold War West Germany and was looking for an adventure. Perhaps he will find more than he expects.

Owen's first job is to get the trucks and earth-moving equipment working. Not much has happened at the site since the last Vixen B test four years previously. Desert conditions and disuse are hard on the gear, and it takes a good two weeks to get the vehicles fit for service. Once everything is moving again, the heavy work begins. Owen operates the Scoopmobile, a front-end loader with a big bucket scoop used to load contaminated Maralinga soil into tip trucks to be carted up to the Taranaki site, where bulldozers will spread it around the blighted test range. The Scoopmobile itself is not allowed to operate in the contaminated areas, because the vehicle will be sold later, and it can't have any radioactivity. The days are unbearably hot for someone from the cold climes of the northern hemisphere.

Owen is part of a crew of 100 men that loads about 76 000 cubic metres of soil each month. They work 12-hour shifts and he knows the work is urgent. The men work at Taranaki, not knowing that the ground is liberally dotted with plutonium-contaminated fragments. No-one says any such thing to him at the time.

The English 'health visitor' Mr Edward Edwards tests material at the site for contamination. Edwards, a friendly and humorous man, calls the lads together when he arrives to give them a lecture. He recounts the difficulties he had back home on the docks in England. To convince sceptical dockers that the low-grade radioactive waste that they had been asked to handle was safe, he had picked up a piece of it and licked it. The Maralinga crew laugh heartily. They can relate to this sort of frontier bravado.

With casual animal cruelty, bored truck drivers mow down sluggish kangaroos that come to Maralinga in the early morning to find moisture, a vision of carnage that will live with Owen for decades. He is especially upset to see joeys run over. Owen also notices that the numerous rabbits in the area are often deformed and have bulging eyes.

After operating the Scoopmobile for a while, Owen shifts up to the contaminated zone at One Tree to do some bulldozing. One Tree is the site of the first major Buffalo test, and there is quite a bit of glazing there. Owen is among the team that performs an emu parade, walking along and picking up the glazed pieces by hand. The health visitor, Edwards, says that the glazing is emitting a little gamma radiation.

At Taranaki Owen operates a bulldozer and also helps to remove the metal bunkers built of steel 2.5 centimetres thick. Later he helps place concrete caps over the pits where much detritus is buried. One of the pits even contains a Canberra aircraft, although this was buried before Owen arrived so he missed the Herculean effort involved in getting the aircraft into the pit.

It's way too hot even for normal clothing, let alone protective gear. Owen discards his long gloves, even though he is picking up glazing known to be emitting gamma rays. He finds his respirator and hat too uncomfortable, so they come off too. The temperature tops 49 degrees Celsius, and sweat pours into the respirator. He can't see and his eyes sting.

At the start he is issued with a minicom – cotton one-piece underwear meant to absorb perspiration. In the early stages of his service he wears a double nylon coverall over the top. The blue film badge he dutifully wears is sent away every month, but he never hears what it says about his exposure to radiation. By the time he gets to One Tree and Taranaki, he wears only shorts, army boots and a white cap. It is the only way to cope with the temperatures.

Every day when he breaks for lunch he wanders over to the mess caravan, operated by Australian civilians, known by the men on site as Queen Mary. Queen Mary herself will later be buried in a pit along with other contaminated vehicles. Owen places his hands inside a hand monitor that is designed to bleep if it detects radiation. It detects radiation nearly every time. He scrubs to remove the traces, but even he – with no prior experience of radioactivity – knows that it is a waste of time since he stays in the same clothing.

Burned out cars, trucks, scrap metal and all sorts of waste including contaminated soil are pushed into the capacious pits at Taranaki, to be covered with sandy soil and concrete. Owen can see the shallow layer of soil placed over the contaminated material is nearly useless. The wind picks it up and swirls it around like whirlwinds he has seen in American movies, 120 metres high and a metre across. Willy willies.

After five months working on Operation Brumby at Maralinga, Owen is discharged from the British Army. Soon after that he notices strange growths on his hands.

In June 1967, the Australian health physics representative JF Richardson visited Maralinga to see how Operation Brumby was going. His brief and remarkably undetailed report was presented to the AWTSC on 17 July. Brumby was still in operation during his visit, so he paid most attention to the visible state of the range, the progress to date and any unexpected problems. He was told during his visit that 633 samples had been collected, including 500 from Taranaki, suggesting that Taranaki was the most problematic site. Despite this, Richardson said of Taranaki, 'A low background [radiation] exists due to induced activity in the soil but no radiological hazard arises for this source'. He witnessed the 30-centimetre slabs covering the pits at Taranaki – 'well-cured concrete resting on the surrounding limestone', each carrying a sign warning of the radiation hazard. He also acknowledged the collection of a large quantity of cobalt-60 beads from the failed cobalt experiment at Tadje. The beads were placed into large lead pots and buried near the airfield cemetery. Richardson's report was among the last formal requirements before the departure of the British from the site. Everything seemed to be in order.

Just as no-one associated with the atomic tests could possibly imagine the specific conditions of the Maralinga lands 24 000 years before them, imagining that far into the future (when only half the radioactivity of the plutonium-239 will have died away) is impossible. McClellan tried to get Pearce to imagine this, but it was a doomed exercise. Whatever use the land will be put to, whatever climatic and geological changes will occur, no-one can predict. As the heavy undertone of the exchanges between McClellan and Pearce attested, leaving the land contaminated was expedient, not prudent. It was not based upon any depth of understanding about the future, or who could be harmed by the radiation now part of the soil and the dust. Geoff Eames, counsel for the Aboriginal people at the Royal Commission, put to Pearce a hypothetical scenario about sprinkling the grounds of Blenheim Palace in England with plutonium in the same concentrations found at Maralinga. Pearce said in this case he would not have recommended the plutonium be ploughed into the earth. It would not be okay at Blenheim, he said, because many more people would be in contact with the soil than at Maralinga. It was essentially a numbers game. And in any case, he said, it was up to the Australians to determine the future use of the range. That was not a matter of concern to the AWRE.

On 23 September 1968, the Australian Government, headed by the Liberal prime minister John Gorton, allowed the British to sign away responsibility for the Maralinga site. The memorandum outlining the terms of the agreement was backdated to 21 December 1967. The agreement was struck on the basis of the flawed Pearce Report and was not seriously challenged by any Australian official until the issue was forced in the early 1990s. In part, the agreement stated that the UK Government was 'released from all liabilities and responsibilities'. In years to come, the British would assert this agreement with considerable vigour and, for quite a while, with notable success. When the British formally withdrew from Maralinga, the federal government assumed responsibility. The plan was eventually to return it to South Australia, although it took longer than originally envisaged.

Did Pearce know that his clean-up operations were inadequate and his infamous report misleading? The question is difficult to answer. At the Royal Commission Pearce vigorously defended his record and stood by his report. He resisted any attempt by lawyers cross-examining him to admit his work was sub-standard or the report was intended to give a false impression. At the same time, he was a British physicist with little real understanding of the differences between the Australian landscape and the gentler fields of England. And he had no real incentive to spend time or effort on divesting the British Government of a site that was by then a white elephant. He was an AWRE insider, so he may well have had access to the data from Roller Coaster that proved to be the key to understanding the true level of Maralinga contamination. However, the Royal Commission did not know about Roller Coaster, and Pearce was not questioned about it. Whether the deficiencies were of commission or omission cannot be determined, although either way the outcome was the same. Moroney defended Pearce at the Royal Commission. Like him, he believed that no-one was likely to come to live at the test site. The whole thing would fade from people's minds leaving a deserted backwater, both physically and metaphorically. Moroney changed his views in the early 1990s, when he calculated the weight of British deceit in terms of kilograms of loose plutonium.

If anything, the clean-up at Monte Bello, which was not subject to an agreement, was even more inadequate than that of the mainland. Some fences and signs were erected at the time, but the islands were not patrolled. The areas were tested in 1962, and radiation levels were found to have fallen somewhat from their 1956 heights. Operation Cool Off in 1965, sparked by news that oil exploration was about to get underway at nearby Barrow Island, involved putting up a few fences around the G1 and G2 sites, but that was about the extent of remediation measures at Monte Bello until much later.

The most concerted effort to bury contaminated debris and remove rubbish was a 1979 Royal Australian Engineers rehabilitation program called Operation Capelin. In 1983, the team from the ARL visited Monte Bello. Geoff Williams and his colleagues found pieces of the frigate HMS *Plym*, including a massive driveshaft (hot with cobalt-60 from the neutron irradiation of the steel), scattered over the beach of Trimouille Island adjacent to where the *Plym* device had been exploded. ARL chief Keith Lokan wrote to the British high commissioner in Canberra pointing out the problem of the British leaving the island contaminated with activated pieces of steel debris from the *Plym*. In response came the put-down

'Everyone knows when you explode a nuclear weapon on a ship, the whole ship is vaporised'. Australian scientists had proof that the ship did not disappear into the ether, in the form of substantial contaminated material lying on the beach. Trimouille was also left with a coating of black dust made up mostly of iron oxides from the *Plym* metal. The 1956 Mosaic tests left about eight islands contaminated.

The tests themselves had many foolhardy elements to them. The clean-up attempts during the 1960s were no better. All three test sites in Australia were left unsafe. By the time Pearce wrote his report and cleared out, no doubt thinking (incorrectly) that his association with Australia was over, probably only the Soviet Union rivalled parts of Australian territory for radioactive contamination. Robert Milliken, writing before the 1990s clean-up, quoted evidence given to the Royal Commission confirming that 'Maralinga is probably the only place in the Western world where plutonium is dispersed without precise knowledge of how much is above and below the ground'. There is no doubt that the British authorities would have been pleased if matters had remained that way. They did not. A great uncovering was about to begin.

10

Media, politics and the Royal Commission

It would seem that rumour, innuendo and conveniently selective recollection place an obligation upon me every six months or so to seek to quieten public agitation which is fomented with respect to the Maralinga tests. The motives of the activists seem, at best, curious.

James Killen, minister for Defence,
House of Representatives, 1978.

I am aware that the British Government and some members of the Australian scientific establishment have adopted the view that this Royal Commission is largely a waste of time.

Justice James McClelland, opening remarks at the
London sessions of the Royal Commission, 1985.

What a difference a generation makes. The great uncovering of the events at Maralinga depended upon factors that now seem inevitable – because they happened. But in fact, the aftermath of the nuclear tests need not have been uncovered at all. The contamination of the test ground might have stayed buried at the site, under layers of secrecy and inertia. This book might not exist but for multiple little uncoverings that brought the saga to light. Journalists played an honourable role, as did nuclear veterans, politicians and Indigenous activists, among others. The secrecy agenda that had seemed so monolithic and immovable at the time of the tests started to crumble in the mid-1970s.

The D-notice system was still in place and the relevant laws had not changed, but Australian politics and media had. In fact, Australian society itself had shifted. In a more complex political situation, the simple truisms of the anti-communist 1950s and 1960s no longer prevailed. The Soviet bloc was not yet dismantled, but, after global progress towards nuclear non-proliferation, the good-versus-evil grand narrative of the Cold War had lost much of its power to animate Australian politics.

The Australian media, having dropped their Menzies era compliance, began nurturing some influential and resourceful investigative journalists who were not interested in comforting the powerful. This transformation of Australian society followed the election, on 5 December 1972, of an ALP government headed by the exhilaratingly reckless Gough Whitlam, a man who crashed through the landscape, exciting some and scaring others. After many years of conservative government, Whitlam threw out the conservative playbook and redesigned the underpinnings of Australian society. A rising generation of ambitious investigative journalists had much to write about, particularly when the governor-general John Kerr sacked the Whitlam government on 11 November 1975, after a period of rapid reform and political scandal. It lasted only three years, but it burned bright and changed everything before it burned out.

The transformation was pervasive. It affected the perception of Maralinga not so much directly, but through the subtle shifts in the public view of government, a casting-off of deference. The way the media treated ministers shifted markedly too. During the 1950s the minister for Supply Howard Beale, the government face of the test program, had informed the media about the tests. This generally involved quarantining some information, and carefully managing the rest. Journalists had largely obeyed his requests to abide by information restrictions, backed by D-notices that outlined specific limits on what the media could report.

Liberal Party Defence minister Jim Killen, who had prime responsibility for managing the issue from 1975 to 1978, had no such control over the media. Initially he seemed caught out about what was at Maralinga. Skilled, motivated journalists kept up the pursuit and placed enormous pressure on him and the government. Killen's *bon vivant* and urbane image, not dissimilar to Beale's, was by this time something of an anachronism. His disdain for media scrutiny worked against him rather than protecting him as it had Beale. The journalists who worked on the Maralinga issue from the mid-1970s had no interest in waiting for carefully crafted and officially cleared media statements. They went looking for their own information.

The development of the more assertive media in Australia was not in itself sufficient to reveal the story about Maralinga's plutonium contamination. The ongoing health problems suffered by both service personnel and Indigenous people who had been in the vicinity of the Maralinga and Emu Field tests led to public campaigns. Also, with the rise of the Indigenous rights movement throughout the 1970s, the prospect of returning the Maralinga lands (including Emu Field) to the traditional owners in 1984 forced discussion on the state of the site into the open.

At first, the issue came to the surface intermittently but quickly died down again. In parliament on 14 September 1972, Lance Barnard, the deputy leader of the Opposition, asked Liberal minister for Supply Vic Garland about radioactive contamination at Maralinga. Barnard specifically inquired as to whether the British had flown in lead-lined boxes of radioactive waste to bury surreptitiously at the test site. Garland's less than satisfactory answer, and a misleading public statement at the same time, came back to haunt Killen a few years later when he initially followed Garland's inaccurate lead. Garland maintained that the radioactive waste buried at Maralinga had a half-life of 15 to 20 years and did not acknowledge the much more dangerous plutonium contamination at the site. Garland had access to the classified Pearce Report, which had been available since 1968 to all security-cleared members of the Australian Government, but his 1972 statements suggest that he had no knowledge of its contents.

In that year, 1972, the French were carrying out atmospheric atomic weapons tests at French Polynesia in the Pacific. This created anxiety in Australia because the tests were so close, and, inevitably, some commentators turned their thoughts to Australia's own role in testing atomic weapons. In June 1972, a story by Michael Symons in the *Sydney Morning Herald*, one of the few that picked up on Garland's statements, harked back to the British test series, quoting Ernest Titterton's response to the airdropped Buffalo shot in October 1956:

> The chairman of the newly formed Atomic Weapons Tests Safety
> Committee, Professor E. W. Titterton, said 'There is no danger
> of significant fallout outside the immediate target area'. That
> was virtually all that was reported in the newspapers at the time,
> although there was a continuing debate among politicians and
> scientists about the Maralinga tests.

The issue rapidly died away. Prominent left-wing ALP politician Tom Uren revived it four years later in federal parliament, when he was deputy leader of the Opposition. Uren was broadly attuned to nuclear issues and had a strong record in opposing uranium mining. His stance was influenced by a series of 1960s articles in *The Age* by Barry Commoner, an American biologist, who had suggested that there was no possible solution to the problem of nuclear waste. Uren was also responsive to representations from nuclear veterans concerned about their health and among the first to place the plutonium legacy at Maralinga onto a crowded political agenda.

Uren asked Killen a bombshell question in parliament on 9 December 1976: 'Is it true that, during the moratorium on nuclear weapons testing between 1958 and 1961, Australia co-operated with the British on conducting secret atomic "trigger" tests at Maralinga and that waste and debris from these tests were buried at Maralinga?' Uren explicitly requested that a Royal Commission be set up to investigate all aspects of the Maralinga test program. He simultaneously issued a public statement saying, 'During [the test] moratorium period the Australian government co-operated with the British government to secretly carry out certain atomic tests in the Maralinga area ... The explosions caused by these tests were so small that they could escape public scrutiny and international detection'.

In his response Killen wrongly described Operation Buffalo in 1956 as the last test series, overlooking Antler, the final major trial in 1957. It was long before the Vixen B tests, too, although he alluded to the minor trials: 'I am not aware of any explosions that took place between 1958 and 1961. I am aware of certain trials, which I distinguish from explosions, as presently advised, that took place. They were conducted pursuant to an agreement between the United Kingdom and Australia'. Killen undertook to carry out further inquiries, while telling parliament he had witnessed the second test in Operation Buffalo. A number of Australian parliamentarians, including Gough Whitlam, had attended this test, having originally been scheduled to attend the first Buffalo test but missing out because of the chaos caused by the delays. Killen even wrote a story for Brisbane's *Sunday Mail*, published on 7 October 1956, titled 'Watched "small" A-blast: sight I will never forget'.

So some tantalising pieces of information were emerging, but the media did not pick up the story. Maralinga was not yet a significant political issue and few people knew the name. The only substantial media references to possible plutonium in the South Australian desert were in the Adelaide papers: the *Advertiser* in a sequence of stories between 3 and 10 December 1976, and the *News* in a prominent article on 17 December.

The first *Advertiser* story, on 3 December, was based on the revelations of Maralinga veteran Avon Hudson, who had been interviewed on the ABC radio program *AM* the previous day. Hudson told the *Advertiser* that plutonium was buried at Maralinga. 'Mr Avon Hudson, of Balaklava, broke 15 years silence last night to talk of his role in what he called a dumping ground for radioactive waste from Britain in the late 1950s and early 1960s.' He asserted that the British has also shipped in radioactive waste to be buried at Maralinga, saying that his conscience had driven him to the media to tell his story: 'It has had a marked effect on my life, knowing there are dangerous elements out there – elements that I now know are the most dangerous things in the world'. These were the same claims that Barnard had asked Garland about in 1972. Hudson told an anecdote about an Aldermaston official who, when he was 'under the weather', asserted that the radioactive contents of barrels that Hudson believed contained plutonium should have been dumped into the Atlantic Ocean in order to save 'a lot of trouble'. While it now appears unlikely that the British actually imported waste to bury at the Maralinga site, Hudson's allegations did direct attention to exactly what was there.

The next day the *Advertiser* quoted a member of a salvage party recovering building materials at Maralinga, Mr E Dutsche, who, citing Defence Department officials, said he had been assured by people at the site that no plutonium had been buried in the area. When Hudson was approached for comment, he said he was not surprised by the denials since this was all he had ever received from politicians. The South Australian minister for Mines and Energy Hugh Hudson (no known relation to Avon) said the issue was a Commonwealth matter. A spokesperson for the Atomic Energy Authority in the UK said 'it was "highly unlikely" Britain had ever exported nuclear waste to Australia'.

Further stories ran in the *Advertiser*. On 9 December, science writer Barry Hailstone brought some scientific fact into the coverage. He reported that Professor HJ De Bruin, who had been a principal research scientist for the Australian Atomic Energy Commission, had called for an inquiry into waste at the site. On 10 December the paper ran a front page story reporting that Killen had ordered an inquiry. The story also reiterated British denials about radioactive waste at Maralinga. It quoted John Coulter, then vice-president of the Australian Conservation Foundation and later an Australian Democrats senator, who said that 'the Australian and British governments had maintained secrecy about nuclear testing at Maralinga after 1957 because such tests would have violated international agreements. But there [still] seems to be a blanket of silence about this'.

Like the stories in *Advertiser*, the front page story in the *News*, which appeared with the huge banner headline 'Plutonium buried at Maralinga', could not have been more prominent. It referred to three reports on the issue of radioactive waste, and although it was not named, one was most likely the Pearce Report, which was still secret. The story indicated that Minister Hudson had called on the federal government to instigate 'radiation monitoring programs for the Maralinga area' and recommended health checks for local Aborigines, a step forwards from his earlier more dismissive position. Still, it seemed everyone wanted to keep their distance. The federal minister for the Environment Kevin Newman advised his department not to get involved, saying that the problem was primarily one for Defence. Foreign Affairs was also concerned about the media interest in Maralinga, especially as Australia pursued its ambition to mine and export uranium.

The claims by Avon Hudson were causing high-level concern. Under Australia's agreement with the IAEA it had to provide an inventory of all fissionable materials and guarantee that no such materials could be used to manufacture weapons. The possibility of this story becoming a serious and potentially damaging problem for the federal government was well known to insiders, including Roy Fernandez, the acting deputy secretary in Foreign Affairs, who pointed out that 'the safety criteria applying 15 to 20 years ago to the storage of plutonium might not be acceptable in the climate of to-day's opinion'. This made 'excessive publicity' and 'unwarranted speculation' about Maralinga undesirable.

In February 1977, Killen, echoing Garland, wrote to Uren saying that there was no evidence to substantiate the claim of plutonium contamination at the site, a position he later had to retract. In fact, around the same time, the *Sydney Morning Herald* contradicted Killen's stance, using his own department's report. The story quoted the government's chief defence scientist Dr John Farrands and claimed that

> Defence Department sources have disclosed that about
> 40 kilograms [almost certainly an over-estimate] of radioactive
> plutonium was buried in a shallow pit at Maralinga, which was
> fenced and guarded for a time. Mr Killen ordered the top-level
> inquiry into tests after the Deputy Opposition Leader, Mr Uren,
> alleged in Parliament on December 9 last year that nuclear
> devices were exploded during an international moratorium on
> nuclear weapons testing.

The story ended with a throw-away line alluding to the fact that unguarded radioactive waste 'might be suitable for use in nuclear weapons'. But other media did not pick this up, at least not yet.

The rumblings continued and a thunderstorm seemed about to break. On 16 February 1977, Whitlam asked Killen if earlier assurances from the British that they had not flown radioactive waste to Australia for burial were still valid. Killen replied that they were. On 31 March, Killen answered a question on notice from Labor MP and later leader Bill Hayden, who had asked on 9 March if he could 'provide a full list of all nuclear explosions which have taken place in Australia giving the date, the size, location and purpose of each?' Killen replied with a basic list of the major trials at the three test sites from October 1952 to October 1957. He included Antler but provided no correction to his earlier suggestion that the major trials had ended in 1956, and he gave no indication that he was aware of the minor trials. Killen then ordered his department to take a closer look. The top-secret Pearce Report that had been in government hands for 10 years showed that plutonium was at the site, even if it underestimated how much was above ground. The full report had not been made public (an edited version was published in May 1979 and tabled in the House of Representatives on 7 June), but it was available to senior government ministers. From the content of public statements by Killen, Garland and others, it appears that few people had read it.

In response to Killen's request, Defence produced the secret Cabinet submission no. 2605, prepared for the Foreign Affairs and Defence Committee and tabled on 11 September 1978. Titled 'Plutonium Buried near Maralinga Airfield', it raised an alarming spectre, that 'a small party of determined men' could recover plutonium 'in a single quick operation if they were willing to take large risks to themselves' and threaten to exploit its 'extremely toxic properties ... against the population of a major city'. In other words, the aftermath of the minor trials at Maralinga could be used for terrorism, a horrifying prospect. Killen wanted permission from Cabinet to seek British co-operation in removing a half-kilogram 'discrete mass' of plutonium buried near the Maralinga airfield. He was not seeking remediation of the plutonium at Taranaki, which the submission conceded was 'practically irrecoverable' because it was so dispersed. (The extent of Vixen B contamination had not yet come out.) The airfield plutonium, noted in the Pearce Report, had been used in a Tims minor trial at TM101 in 1961.

Members of the Foreign Affairs and Defence Committee accepted the recommendation of the submission that the government should avoid public scrutiny and actively attempt to limit media knowledge. The committee specifically asked Killen to 'arrange for a reconnaissance to determine what would be entailed in exhuming the discrete mass of recoverable plutonium buried at Maralinga on an assumption that it would then be removed from Australia by the British'. The submission advised against announcing the purpose for the reconnaissance, suggesting, if necessary, that it be described as a 'review of physical security measures and possible need for maintenance work'.

The committee acknowledged the need for a public statement 'about the exhumation/repatriation operation', but the 'timing and text would require discussion with the British'. The long-established method of stonewalling media scrutiny, or diverting attention with a bland and uninformative public statement, still operated, but not for much longer.

The secret submission also noted that the Maralinga plutonium might compromise Australia's international obligations. The prevailing international agreement was the Treaty on the Non-Proliferation of Nuclear Weapons, also known as the Non-Proliferation Treaty. Since 'some of the nuclear material buried at Maralinga may be safeguardable', this meant it had to be declared 'to the International Atomic Energy Agency (IAEA) on Australia's inventory of materials'. The problem wouldn't go away by itself: 'The IAEA is aware of the possible presence of undeclared safeguardable material at Maralinga'.

The submission put forwards three options for dealing with the plutonium. It favoured asking the British to dig it up and take it home, which was, eventually, enforced. The Foreign Affairs and Defence Committee formally responded to Killen's submission on 28 September 1978 confirming that it would approach the British. The nuclear affairs division of the Department of Foreign Affairs found that the plutonium in question was buried in six containers in 'relatively shallow' pits. Despite the term discrete mass, it was not a single lump but dispersed through salt packed into the containers. The British, after initially refusing, recovered the Tims plutonium in 1979 and took it back to Britain.

Despite their desire to keep the story out of the public domain, Killen and his department reckoned without the efforts of one of Australia's leading investigative journalists. Brian Toohey, then 33 years old, had been a political correspondent for the *Australian Financial Review* since 1973 and later became Washington correspondent and then editor of the *National Times*. He was well known for his unwillingness to stick to official versions of events. Fellow political correspondent Mungo MacCallum noted in a 1989 feature, 'As a journalist of unrivalled application and extraordinary contacts, Toohey dedicated himself to opening up things that politicians (and more particularly bureaucrats) wanted to keep secret, irrespective of what they were'. Fatefully, Toohey got his hands on the secret Defence submission. While he had to maintain confidentiality, he identified his source as 'someone in the government who thought the information should be public, without being motivated by either a strong environmental, or nuclear disarmament, perspective'.

Finally the ticking time bomb of Maralinga was revealed to a large audience. Toohey used the leaked Cabinet submission as the basis of his story in the *Australian Financial Review* on 5 October 1978, under the headline 'Killen warns on plutonium pile'. The sub-heading was even more compelling: 'Terrorist threat to British atomic waste'. It was a factual account of the contents of the submission, with commentary on the consequences. For example, the second option outlined by the submission for dealing with the plutonium involved Australian authorities extracting and analysing the plutonium at the site, a task described in the submission as potentially beyond the limits of Australia's capacity to deal with radioactive materials. 'Although the submission does not make this point, the public is hardly likely to be reassured by the revelation that despite all the money spent on nuclear research within Australia, half a kilogram of plutonium is possibly too hot for the authorities to handle in line with IAEA requirements.' Toohey noted that the Menzies government in particular had not demanded sufficient safeguards at Maralinga: 'The submission ... makes clear that Australian Governments in the past have taken an extremely lenient attitude towards the existence of the Maralinga plutonium through its nuclear weapons tests in Australia in the 1950s'. He concluded, 'It is now 20 years since the tests finished. The fall-out, however, is still a very live issue in British–Australian relations however much both Governments want to keep the negotiations entailed in last Thursday's Cabinet submission a closely guarded secret'.

Over 30 years later, Toohey viewed the story as part of a continuum that forced government accountability: 'These articles were worth doing because they gave the public a glimpse of what was being withheld in a democratic society'.

Toohey's revelations were quickly picked up. The *Sydney Morning Herald* assigned reporters to travel to Maralinga and see what was there. Killen had said that, thanks to the Toohey article, he had to urgently upgrade security at the site. 'When journalists flew in there was no sign of the increased security measures announced on Thursday night to guard a buried lump of plutonium from terrorists ... When told of this yesterday the guards at Maralinga just chuckled', the *Herald* reported on 7 October.

In the same issue a familiar name reappeared. Ernest Titterton wrote a feature-sized spread to answer the growing controversy, his final free kick before the McClelland Royal Commission robbed him of credibility. The feature conveyed an impatient tone that Titterton undoubtedly felt at having the plutonium issue dredged up after all these years. The media coverage since Toohey's story was 'near to hysterical', he claimed, but the buried plutonium didn't pose any danger. Someone could carry it around in their pocket and no harm would befall them. There was limited truth in this, although Titterton did not explain how dangerous even a tiny particle could be if it was inhaled or ingested. He was on even shakier ground in claiming that terrorists would have no use for half a kilogram of plutonium, since all plutonium could potentially be useful to them. He argued that the British would hardly leave buried a quantity of plutonium, a valuable material that might be worth tens of thousands of dollars, if it were in any sort of usable form. He also made claims, later shown to be wrong, that Brumby, the 1967 clean-up, had taken care of these problems. He insisted, 'Putting aside politics and emotional grandstanding, it is clear ... that the public need have no worries about terrorist activity'.

The *Herald*'s political correspondent Peter Bowers examined the mystery of the buried plutonium on the same day. Bowers wrote that Uren was calling for Killen's resignation in light of the recent revelations about minor trial contamination from William (now Lord) Penney. For the first time, Penney had revealed some information about the minor trials, particularly the Tims trials that had resulted in the infamous 'discrete mass' of plutonium. This was the beginning of revelations about the residue of the minor trials: 'Lord Penney revealed that small-scale nuclear tests, which he described as "little mock explosions", were conducted apparently long after full-scale bomb testing ceased. The experiments have remained a highly classified secret for the past 17 or 18 years'. Uren suggested that Killen had misled parliament by denying what was left behind at Maralinga.

Two days earlier, on 10 October, the *Herald* had reported that Britain had been formally asked to repatriate the plutonium to the UK. The speed of this request appeared to reflect media pressure, with the story noting 'that the Government had been forced to act quickly' after the Toohey story's published details of the Cabinet submission. A supplementary piece reported the official British stance that lasted until the McClelland Royal Commission. A spokesperson for the UK Atomic Energy Authority said that, 'although his department had no record of what was left at Maralinga … it was unlikely that any plutonium was involved'. He also denied that any atomic waste had been sent from Britain to Australia.

This was the same official line run back in 1976 in response to the Adelaide stories. By the next day, 11 October 1978, as pressure mounted, the *Herald* reported that the British were planning to send a team of experts to Maralinga to investigate the remaining plutonium as both governments tried to damp down concern. The British high commissioner Donald Tebbit dismissed the idea that the plutonium could fall into the hands of terrorists: 'Even the [Great Train] robbers would have trouble coping with this situation. They might do better with a toy pistol', he said. This story also gave more detail about the much-discussed 'discrete mass', which had dominated media coverage since 5 October, with Tebbit explaining 'this material originated in six separate minor experiments' when a small disc of plutonium was shattered into numerous fragments 'which were collected into a steel container filled with common salt. No nuclear explosion was involved'.

Another original Maralinga participant emerged back into the light. On 11 October the *Herald* quoted Howard Beale, then in retirement, who dismissed as ridiculous any claims that the buried plutonium could be a terrorist target. He reserved a portion of his scorn for the source of the story: 'What right has an official in the government to play God and leak documents of this nature? I think it was immoral and quite wrong to let this document loose'. Beale lectured reporters on their particular responsibility to assess the national interest before publishing and, perhaps reflecting his own approach, also remarked on the procedures that should have been in place to prevent such a leak: 'An issue as sensitive as the Maralinga one should have been handled by the smallest number of people possible'.

Toohey's follow-up story on 11 October added more fuel. This article, titled 'Maralinga: the "do nothing" solution', brought the wrath of Killen down on Toohey's head. The story questioned the Australian Government's response in light of a statement issued by the British High Commission on 10 October that nothing needed to be done. The story quoted a radio interview in which the acting Australian Foreign minister Ian Sinclair agreed with this and said that the plutonium was safe where it was buried. The extent of tensions between the Australian and British governments did not come out in the story, however. Behind the scenes, confidential cables were being exchanged between the UK deputy high commissioner in Canberra Henry Dudgeon and the Foreign and Commonwealth Office that belied the apparent agreement of the public pronouncements. Dudgeon mentioned the 'cavalier tone' of the official Australian request for the removal of the plutonium and declared that 'the Australians of course fully accept that we no longer have any legal obligation relating to Maralinga'. Toohey was not deterred when Sinclair backed the British stance with a supportive statement that played down the risks, particularly as the Cabinet submission had made strong statements about the terrorist threat posed by the material at Maralinga.

Toohey's follow-up story also picked up on Jim Killen's responses to the original one:

Mr Killen relied on his verbal dexterity in several answers he gave in Parliament yesterday. For example, he said that no Cabinet submission prepared by himself had said the plutonium at Maralinga was 'currently' a terrorist threat. Last Thursday he said in his press statement that the [*Australian Financial Review*'s] report of his submission and its emphasis on the potential terrorist threat provided sufficient cause for him to substantially increase security at Maralinga on that very day.

The story reported that Killen had told parliament he found out about the plutonium problem only in early 1977, which seemed to accord with his public statements. In 1976, when Tom Uren and some elements of the media had begun to question what was at Maralinga, the Defence minister had seemed not to know and had asked his department to dig deeper. Toohey concluded his contentious 11 October story by tying it to a then-current political debate: 'If the Government ends up doing nothing about the Maralinga plutonium it will only have succeeded in raising public doubts about the safety of nuclear materials at the same time as it is trying to convince the world that any Australian uranium exports will be on the strictest possible safety terms'.

Upon publication of this second story, Killen denounced Toohey in parliament. He accused Toohey and the *Financial Review* of issuing an open invitation to terrorists to help themselves to the dangerous material at Maralinga, and of reporting falsehoods. He said, 'It is a day for regret when a journalist and a newspaper, aided by a criminal act, have published a story that is against the interest of the nation and its people'. Killen's outburst in parliament was reported in the stablemate Fairfax broadsheet the *Sydney Morning Herald* on 12 October 1978:

> [Killen] said a report in the *Financial Review* 'written by one of that paper's employees' has stated that he suggested there might be no need to do anything other than upgrade the police guard at Maralinga. 'I said no such thing and suggested no such thing', Mr Killen said ... 'This is a pernicious, wicked and odious technique that has long been practised by this man', Mr Killen said ... 'The person concerned with the report wouldn't be capable of accurately reporting a minute's silence', Mr Killen said.

Killen claimed that the Defence submission did not make assertions about an immediate terrorist threat, although one could conceivably exist if no action were taken, and that publicising this threat 'was an act of irresponsibility'. Toohey took the attack in his stride: 'I knew that I had accurately reported Killen's cabinet submission, despite his flamboyant, often incomprehensible, accusations against me. Perhaps his reaction reflected the way he was unaccustomed to media criticism'. When he was interviewed in 1987, Toohey nominated his Maralinga plutonium story as a career highlight:

> Jim Killen went berserk when [the discrete mass of plutonium] was revealed and he banned the paper from any contact whatsoever with the Defence Department. [Malcolm] Fraser ordered him to lift the ban, but what Killen went ape about was the story being a breach of security. My point was that the real security problem lay in leaving unguarded plutonium at Maralinga.

While the drama around Brian Toohey's plutonium disclosure was unfolding, the legendary denizen of the Canberra press gallery Mungo MacCallum watched with wry amusement. MacCallum, a colourful satirist and political correspondent, spent 20 years in the federal parliamentary press gallery. His column 'From the gallery' of 12 October 1978, titled 'Killen throws a Maralinga bomb – with fallout', chronicled Killen's lambasting of Toohey:

> Mr Killen exploded in the megaton range, and scattered
> his fallout widely – especially over this paper's Canberra
> correspondent, Mr Brian Toohey. Mr Killen never actually
> named Mr Toohey; however, in a series of answers to questions,
> and in a ministerial statement designed to clear up the whole
> issue, he left no doubt as to his primary target.

MacCallum recounted how Opposition leader Bill Hayden revealed that he had contacted Whitlam, and the two Whitlam government Defence ministers, Lance Barnard and Bill Morrison, who had said they did not recall hearing about plutonium buried at Maralinga. Killen countered this with the fact that the still-classified Pearce Report, which noted the Maralinga plutonium, had been available to them when they were in government. This was possibly a self-defeating point to make, given Killen had also expressed ignorance about the Maralinga plutonium before 1977 when he, too, had access to the report. MacCallum said that Killen, having savaged Toohey, 'sat down to a big laugh and a round of applause'.

Toohey prepared two more stories in this series. On 12 October he quoted Killen saying, 'It is characteristic of a certain kind of so-called journalism in this country that certain sections of my Cabinet submission were reported accurately, while other parts were selected for distortion to contrive a mixture that would create a sensational impact and alarm the public'. The next day Toohey questioned whether the size and unwieldiness of the Defence Department bureaucracy had contributed to the Maralinga plutonium controversy: 'In the Maralinga case … there have been accusations within the department that relevant information about the plutonium buried at the South Australian site of the British nuclear tests has not flowed to all levels of the department that needed to know'.

By 13 October 1978, both government and Opposition politicians were claiming that they were ignorant of what was at Maralinga, that others had misled them, or that someone from the other side had misled parliament. Hayden said Garland had misled parliament in 1972, while former Labor Defence minister Lance Barnard said his own department had misled him about Maralinga in 1973. The former Labor minister for Environment and Conservation Moss Cass said that he could not remember being told about plutonium buried at Maralinga, a claim undermined by a letter he wrote dated 3 December 1974 in which he referred to 'long lived and highly radioactive wastes contained in the Airfield Cemetery', seemingly a reference to the 'discrete mass' later removed by the British. The letter did not specifically mention plutonium but did mention contamination both at the airfield and at Taranaki and called for a survey to better understand how radioactive wastes had been stored, dispersed and taken up by the 'biosystem'. Prime Minister Malcolm Fraser countered by saying that the previous government had had as much information available as the current one. While this claim was true, it did not reflect much credit on the efforts of his own ministers.

Reacting to media reports about Maralinga and concerns from his state constituency, the premier of South Australia Don Dunstan wrote to Fraser asking for a full inquiry. The letter was tabled in evidence to Senate Estimates on 17 October 1978 and contained the following statement: 'On a matter of such fundamental significance to public health and safety as the proper disposal of plutonium and other high level radio active wastes, it is essential that the fullest information on security and other precautions be assembled'.

As the story reverberated around Canberra, Fraser asked for briefings. A senior adviser in the Resources Branch, GF Cadogan-Cowper, set out the history of the issue in parliament from 1972 in a briefing note dated 13 October. He noted that the Opposition might focus on Vic Garland's answer: 'Should you be questioned on the dumping of wastes it may be necessary to note that the advice Mr Garland received was apparently incomplete'. In questionable advice, given the long half-life of the Maralinga plutonium, he told Fraser, 'Emphasis could be laid on the short half life of the fission products and that because of their short half life the quantity has decreased rapidly over the 20 years since the tests'.

A few days later in the *Herald*, Peter Bowers took stock of the frenetic activity since the Toohey story had broken: 'We have learned more about what is buried at Maralinga in the past week than in the past 20 years. And there is much more yet to be learned about the Maralinga caper'. The era of revelations was now underway. Bowers summed up his view of the events:

> The real issue is why the presence of plutonium had been kept
> so long not only from the Australian people but, apparently, from
> the Australian Government. The real danger – the ever present
> danger – is that governments and their bureaucracies are secretive
> and tell the public only what they think the public should know.
> The Australian public would still be ignorant of what was buried
> by the British at Maralinga 20 years ago [were] it not for the fact
> that a Cabinet document was leaked to a reporter.

Brian Toohey set in motion years of media scrutiny of the legacy of Maralinga. He maintained an interest in the story well into the term of Bob Hawke's Labor government that came to power in March 1983. By then Toohey had moved to the *National Times*. Several months before the McClelland Royal Commission into the British nuclear tests began, Toohey wrote a feature based on another leak titled 'Plutonium on the wind: the terrible legacy of Maralinga', which contained a detailed examination of the Vixen B issue. Toohey had obtained the full, uncensored Pearce Report, still classified at that time and available publicly only in what he called a 'sanitised' form. He was unable to reveal his source but said, 'The backdrop was a concern that a proper clean-up occur'.

The *National Times* feature had much more to say about the nature of Vixen B than his earlier stories: 'The experiments were usually described as point safety tests, despite the obvious irony in the use of the word "safety" for operations that left plutonium scattered across the countryside'. The feature was an indictment of the Maralinga plutonium legacy. 'It would seem that what the British and Australian authorities described as minor experiments in fact involved the cavalier dispersal of plutonium and have created a far greater health hazard at Maralinga than the full-scale atomic tests.'

Most mainstream news media began reporting the Maralinga story after Toohey's articles in the *Australian Financial Review*. One political casualty of the tumult was soon apparent when Cabinet moved responsibility for Maralinga from Killen's Defence portfolio to the minister for National Development Kevin Newman. 'This follows a row which highly embarrassed the government over the deposits of plutonium at the toxic waste site at Maralinga in South Australia', *The Australian* reported on 10 November 1978.

The removal of Jim Killen did not slow the story down. *The Australian* kept up pressure on the federal government with a front page story featuring the huge headline '"Take it back" requests ignored: British snub on plutonium plea'. This story highlighted a problem that dated back to the time of the tests themselves – that the British were slow to answer an Australian request.

> The Government realises that it must take some action over the 'recoverable' plutonium because of its obligations under international nuclear safeguard agreements which have a strong bearing on the future development of the Australian uranium industry. It is understood that the government decision that the plutonium should be removed has received a sympathetic response from bureaucrats in England but this has not been matched by the response of the politicians.

This story prompted more activity in federal parliament. Labor senator Gareth Evans drafted a question to the leader of the government in the Senate John Carrick, a rough draft of which, with short-hand forms of expression, remains on the official Maralinga file:

> Is it true as reported in this morning's *Australian* that the British govt has snubbed Ausn requests to remove waste plutonium buried at Maralinga ... in that it has failed to respond to Ausn requests to this effect by the required deadline of 7 November? If this is so, and if the British govt continues to remain unbeguiled by the subtleties of Ausn diplomacy, what other plans does the Ausn Govt have in mind for the safeguarding of this material?

Senator Carrick prepared a reply claiming he hadn't read the story and had no knowledge of any breakdown in discussions between the Australian and British governments. The reply belied the activity behind the scenes. The Department of Foreign Affairs sent a cablegram to its London officials summarising the substance of *The Australian* story. This followed up a cablegram five days earlier, in which the growing crisis was spelled out:

> Ministers remain under considerable pressure on this issue from press and parliamentary questioning ... In addition to questions without notice, some fourteen questions on Maralinga and the visit of the British technical team are on notice to be answered. There is also press speculation that we have to rely on Britain on the alleged ground that AAEC [Australian Atomic Energy Commission] is not capable of dealing with the problem.

Finally the pressure was relieved when the federal government extracted an undertaking from the UK to remove the airfield plutonium. It was a win. Malcolm Fraser wrote to Premier Dunstan in February 1979, saying, 'I am glad that a result so satisfactory to both our Governments has been achieved'. Fraser also became more comfortable about Maralinga information being released to the public. He said in this letter that 'as much information on Maralinga as possible should be made public'. He even recommended the release of most of the discredited Pearce Report, minus details about the locations of buried contaminated materials. It was months after Brian Toohey had made the main information in the report public.

The *Advertiser* began a high-profile campaign seeking justice for the nuclear veterans, resulting in a series of stories run over a week in April 1980. The stories, by reporters David English and Peter De Ionno, presented case studies backing the calls for compensation for service personnel said to have been harmed by their service at Maralinga. The stories were bolstered by an editorial on 17 April 1980:

> The testing of British atomic weapons at Maralinga … ended
> many years ago, but the consequences linger on. There was
> a brief flurry in 1978 when it was revealed that potentially
> radioactive waste material, since removed to the UK, had
> been left at the test site. Now there is further, and more
> serious, concern at the disclosure of the possible effects of
> radiation contamination of people exposed to the fall-out
> from those tests.

These stories began putting names and faces to the statistics of service personnel who had been at Maralinga. The Irish immigrant James Barry had died of cancer in 1966 at the age of 50 after working as a builder at the test site. His photo appeared under the heading 'A victim of Maralinga?' alongside a picture of his widow. The story claimed that about 20 ex-service personnel had died of cancer or had contracted it. It quoted Barry's widow, Mary Jane Barry: 'He wasn't supposed to tell me anything, because of the Secrets Act and all, but he told me bits and pieces. He said that things were very lax up there; they didn't take enough precautions'.

The *Advertiser* series gave a forum to prominent aggrieved ex-Maralinga hands such as Avon Hudson in more detail than ever before. In one article, Hudson addressed the possible breaching of international agreements: 'Mr Hudson believes that nuclear bomb tests were conducted by the British on the range after the [official] bomb-tests. Atomic weapons tests after 1958 would have been in breach of an informal moratorium on bomb experiments made between the UK, US and USSR in 1958'. The same article mentioned Maralinga veteran Richard (Ric) Johnstone, one of the first (in 1973) to receive a Commonwealth pension when unable to work because of symptoms, he said, that were due to his six months of service at Maralinga during the 1956 Buffalo series. In 1988, he was the first person to be awarded damages by the courts, after a long battle, winning $679 500 in compensation.

In April 1984, the Australian magazine *New Journalist* ran a critique of the test era journalism. In 'Buffalo Bill and the Maralingers', Lindy Woodward was scathing of their role. Journalists had, as now, an important role in deciphering pronouncements on the safety of atomic technology, she wrote, but instead took 'the experts' at their word that the tests were totally safe and crucial to peace. 'It was a national suspension of disbelief, indulged in and encouraged by the media.' Even British journalist Chapman Pincher, seen as a troublemaker and 'scoop journalist' by the Australian and UK governments at the time, fell short in this account: 'Chapman Pincher, the science writer from the London *Daily Express*, was the *Advertiser*'s own "expert" on the tests, but his reports were short on scientific analysis, and big on British enthusiasm for what was going on in the Australian desert'.

A new federal government came to office on 5 March 1983 under the leadership of Labor's Bob Hawke. For its first 18 months, Senator Peter Walsh as minister for Mines and Energy was responsible for dealing with the Maralinga aftermath. Walsh was widely disliked, a dour, 'dry' economic rationalist, with little of the urbane charm of Beale or Killen. He soon understood the issue of British nuclear tests could become a major political problem for the new government if it was not dealt with expeditiously. During 1984, he issued prolific media releases on the tests, until a reshuffle late in the year saw him head off to a legendary stint as Finance minister. After Walsh's departure, Gareth Evans took on the portfolio.

Walsh commissioned the Kerr Report into the risks to the Australian population from atmospheric fallout during the tests. He announced the report, under Charles Kerr, professor of preventive and social medicine at the University of Sydney, on 15 May 1984, and it reported 16 days later. It was not a public inquiry, but Kerr did have powers to call expert witnesses and to examine all published scientific literature and other data relevant to the tests. Kerr's report forcefully criticised the most comprehensive account of the British nuclear tests to that point, the 1983 AIRAC 9 report, commissioned in 1980 by the minister for Science and the Environment David Thomson. The McClelland Royal Commission subsequently endorsed the demolition job carried out by Kerr.

AIRAC fought back. Emeritus Professor AM Clark, its chair in 1984, wrote to Barry Cohen, minister for Home Affairs and Environment, saying the Kerr Report was not objective and contained numerous 'internal contradictions and apparent misunderstandings'. Clark's objections came to nothing – AIRAC, with its roots in the AWTSC, was discredited. Kerr called for a public inquiry to further probe the serious issues that his team had turned up. Walsh, although apparently opposed to setting up an expensive inquiry (according to radiation scientist Peter Burns, Walsh 'said the Royal Commission was just a lawyer's picnic, a waste of time and money'), was forced into it as the weight of evidence became too heavy and the political risk too great. The ARL scientific team in 1984 came back with evidence of loose plutonium in large quantities on the site, prompting Walsh to make statements about the need to fully understand exactly what was there. The fundamental disagreement between AIRAC and the Kerr Report was the final straw.

Walsh announced the establishment of the Royal Commission on 5 July 1984. In his media release, he indicated that the inquiry had been charged in particular with examining 'measures that were taken for protection of persons against the harmful effects of ionising radiation and the dispersal of radioactive substances and toxic materials as judged against standards applicable at the time and with reference to standards of today'. The Royal Commission was headed by Jim McClelland, a colourful former Whitlam government minister known to many as Diamond Jim. At that time, McClelland was chief judge of the Land and Environment Court of New South Wales. The two other commissioners were Jill Fitch, senior health physicist for the South Australian Health Commission, and William Jonas, lecturer in geography at the University of Newcastle.

By the time he wrote his autobiography, Walsh had major regrets about the McClelland Royal Commission, labelling it as 'the most unambiguous mistake I made in Government. As is usual with Royal Commissions, the terms of reference were stretched, the budget blew out and the reporting date extended ... Lots of us approved when Jim McClelland tipped buckets on Menzies, but this did not justify the $3.5 million it cost the taxpayers'. However, at the time, he showed public support. He had little choice, particularly given the plutonium uncovered at the site in May 1984.

The McClelland Royal Commission was officially opened in Sydney on 22 August 1984, with a second formal opening in Adelaide on 11 September. Oral evidence was taken in Sydney, Brisbane, Melbourne, Adelaide, London and Perth, as well as in remote locations at Marla Bore, Wallatinna and Maralinga in South Australia, and Karratha in Western Australia. After 116 sitting days, all in open session, the final sitting was supposed to be on 26 July 1985, although another sitting was needed in September to hear final submissions. The Royal Commission took oral evidence from 311 witnesses, including 48 Aboriginal people, 18 Australian scientists or technicians and 241 Australian service personnel. It travelled to London to interview a full roster of British witnesses, 40 in all. William Penney, the star witness, was subjected to questioning that was much more probing than any he had been exposed to in the 1950s; enduring it must have been difficult for someone more accustomed to keeping his own counsel. McClelland actually liked Penney, describing him in a later interview as 'a nice old man. I got the impression that he wasn't terribly proud of having used his immense scientific skills on an exercise that was really an exercise in futility'. McClelland thanked Penney warmly for his evidence, given while the old nuclear scientist was ailing, and singled him out in the official Royal Commission acknowledgments, saying, 'In particular, Lord Penney, who interrupted his retirement to give the Royal Commission the benefit of his unique experience and vast knowledge'.

The epic transcript contains the historical and technical story of the test series, but also the human story. In many ways it is a remarkable document, with moments of tears, sadness, humour, frustration and fury. The final report strongly condemned just about every aspect of the British nuclear tests. It was long-delayed national revenge, needed to lance a boil. Australia has held scores of Royal commissions over more than a century. Few have been as thorough, angry or applauded as this one. Not all of its recommendations were achieved, however. First, the recommendation that a Maralinga commission be established to oversee a clean-up and manage the range, with representation from the traditional owners as well as the UK, Australian and South Australian governments, did not come to pass in the form envisaged by Jim McClelland. However, the Maralinga Tjarutja people were represented on the Maralinga Consultative Group with representatives from the South Australian and UK governments. Also, the Australian Government paid for independent scientific advice for the Maralinga Tjarutja people beyond the recommendations of the Royal Commission report. Second, no national register of Indigenous people and veterans harmed by the tests was ever established. Third, despite the report's recommendation that the UK Government pick up the tab for the clean-up, after years of wrangling, Britain paid less than half. The real power of the

Royal Commission was in the fact that it took the side of Australia against its nuclear coloniser.

During the months that the Royal Commission was taking evidence, the nuclear tests maintained a high profile throughout the mainstream media, and many publications assigned reporters to attend the hearings. Journalists Paul Malone and Howard Conkey in a feature for the *Canberra Times* asked Ernest Titterton what was known about the plutonium contamination risks at the time of the experiments. In his typically cantankerous manner, referring to the 12 Vixen B tests, he asked them, 'Wouldn't you expect plutonium around the place? Of course there is plutonium around the place, it is always there, it was always expected'. He said that it had not been possible to go around and pick up every fragment of plutonium, some of which was not sufficiently radioactive to be reported. He maintained his nuclear warrior persona until the end, even as the truth about Maralinga was uncovered at last for the Australian public to judge.

1 1

The Roller Coaster
investigation

*The results from Roller Coaster show that the measurements
made in the Vixen B trials underestimated the ground deposit
of Pu by a factor of ~10 ... Obviously this was known at
Aldermaston in 1966 when the program for cleaning up Taranaki
was being developed, but it was not conveyed to Australia.*

John Moroney, unpublished aide memoire, 1992.

*I was really angry when I read all the reports, then we got out there
and suddenly found ourselves knee-deep in plutonium.*

Peter Burns, part of the 1984 team of radiation specialists
who surveyed the Maralinga test site, 2004.

We may never know the intent, but we do know the consequences.

Ian Anderson, on the *Science Show*,
ABC Radio National, 1993.

Maralinga was uncovered incrementally. Bit by bit, as worrying
details came to light, Australia's citizens learned what had
been done a generation earlier. It did not all come out at the Royal

Commission, either. The last piece of the jigsaw – the extent of the contamination, and the fact that the British did not share what they knew about it – took a concerted effort by John Moroney, a once-loyal servant of the British tests who analysed the data, and a dogged journalist, Ian Anderson, who brought it to public attention.

Discovering the true nature of the plutonium tests at Maralinga was a significant and celebrated achievement. It required sophisti-cated journalism informed by a strong understanding of science and an ability to get at hidden information and make sense of it. Ander-son's story, 'Britain's dirty deeds at Maralinga', which appeared in the British-based weekly magazine *New Scientist* on 12 June 1993, was considered to be a landmark story by both the experts who provided its source material and the peers of the journalist who wrote it. More than ever before, the story revealed the truth of the minor trials.

Anderson was Australian editor of *New Scientist* and a pioneer-ing science journalist, loved and respected among his peers. His work on Maralinga was part of an enviable legacy. He was a leading contributor to the development of Australian science journalism, through his position at *New Scientist*, his foundation and leadership roles in the main professional organisation for science journalists and communicators, Australian Science Communicators, and in his initiation of ScienceNOW, the festival of new science held in Mel-bourne. In Anderson's obituary in *New Scientist* in 2000 (he died prematurely at the age of 53), his close friend Tim Thwaites wrote, 'No-one has contributed more than Ian to the promotion of Aus-tralian science and technology to the world. Through the excellence of his reports in *New Scientist* and other publications, he presented Australian research to an international readership'.

'Britain's dirty deeds' earned Anderson two Michael Daley awards for science journalism and appears to have influenced the course of ministerial talks when Australia was negotiating with the UK for a monetary contribution to help clean up the Maralinga site. This story marked the first time that the extent of plutonium contamination at the desert test range was made public. It also exposed the fact that the British authorities had known the level of contamination and covered it up. The story contributed 'moral pressure' at a crucial moment, opening disturbing new information to public debate that raised fundamental questions about the nature of the Australia–UK relationship. Anderson's story appeared at a crossroads moment in the history of the toxic old site. The story triggered a renewed media interest in this particular example of nuclear colonialism.

Anderson had a ready analogy when he discussed his story. When it was published, the Australian cricket team was playing at Lords in London in an Ashes series. He told listeners on several radio shows that Australia was facing 'the old enemy in another arena'. The theme of ongoing battle between traditional adversaries was especially resonant since Anderson published his bombshell in a popular British publication with a large British readership. An editorial in the same edition substantively supported Anderson's story.

As the Australian cricket team faced the bowling attack at Lords, Australia's Energy minister Simon Crean and Foreign minister Gareth Evans were entering the finale of a long-running dispute to obtain funds for a large-scale clean-up operation at Maralinga. Archie Hamilton, the Conservative British minister of state for the Armed Forces, and others in John Major's government, argued that their responsibility had been signed away in the 1968 joint agreement. Anderson's story asserted that the Australian Government, when it signed this agreement, was not informed that the British test authorities knew Maralinga would remain toxic for tens of thousands of years into the future.

Like many influential stories, 'Britain's dirty deeds' had a serendipitous beginning. Anderson took his car to Heidelberg Mitsubishi in Melbourne's northeastern suburbs one day in early 1993. Scientist Geoff Williams, whose car was also being serviced, ran into Anderson in the waiting room. Williams was one of the small team of radiation specialists who had gone to Maralinga in 1984, the expedition that had led directly to the establishment of the McClelland Royal Commission later that year.

At the Heidelberg garage, Williams and Anderson, who knew each other from some long-forgotten *New Scientist* story, began chatting about Maralinga. Anderson had reported on Maralinga a few years previously, and Williams knew enough of Anderson's work to trust he would understand the complicated material he discussed. He told him about some new safety trial data from the US that had implications for understanding the plutonium contamination at Maralinga. By the time Anderson drove his freshly serviced Mitsubishi home, he had the beginnings of the story.

Williams had suggested that Anderson contact his colleague and senior ARL manager John Moroney, who had inside information. Moroney had been studying for a masters degree in physics at Melbourne University under Professor Leslie Martin in 1957, when Martin was about to retire as chair of the Australian AWTSC. Moroney, considered an efficient, intelligent and scientifically literate young man, gave up his studies to join the AWTSC as secretary. He stayed until it disbanded in July 1973, becoming one of the world's leading authorities on atomic fallout in the process. When Williams suggested Anderson talk to him in 1993, Moroney was head of the Radioactivity Section at ARL.

Several years before, the Royal Commission had provided a mechanism to call witnesses, review thousands of pages of documentary evidence and breach the persistent secrecy around the British tests. However, exhaustive as it was, it could not tell the whole story. In the early 1990s, ARL obtained newly declassified documents about radiological tests in the US known as Roller Coaster. The Royal Commission did not consider the Roller Coaster documents because they were not available when it sat.

The Roller Coaster tests had gone virtually unnoticed outside Nevada, and were not listed as part of the publicly available record at that time. Hardly anyone knows the name even now, and the tests are not mentioned in books about Maralinga. When the ARL scientists, and specifically John Moroney, started to look through the records they began to understand that radioactive contamination left behind at Maralinga was far greater than they had believed before 1984 when the Pearce Report was still considered accurate. The subsequent Royal Commission, which owed much to the astonishing revelations of the landmark 1984 site expedition, received precious few British records about Vixen B, so while it did note Vixen B left significant contamination, the true extent of it remained unknown until Moroney's later investigation.

Moroney was one of the few to recognise the significance of Roller Coaster, joint US–UK tests that were similar to Vixen B. The main difference was that during the Roller Coaster experiments a much greater effort had been made to accurately measure the level of plutonium contamination. The thoroughness of the Roller Coaster documents contrasted with the paucity of information for Vixen B – at least, the information made available to Australia. Moroney described the Roller Coaster data: 'Typically, they are thorough records of the Operation, going into infinite detail'.

Moroney provided the data that made the *New Scientist* story possible, based on documents from the National Technical Information Service in Washington DC. He had been through each and every page of these mostly microfiche documents, recording and analysing as he went. Moroney analysed about 2500 pages of declassified nuclear contamination data from the US–UK Roller Coaster trials in Nevada (held in May and June of 1963, after the British had resumed weapons testing with the Americans). While some details of the tests remained classified (as Moroney said in a letter to Pat Davoren, 'There are still wraps around the main core of Roller Coaster data, with a few bits sticking out for mere mortals like us to see'), he had enough to understand their significance. The Roller Coaster data enabled Moroney to make a detailed scientific case that the British atomic test authorities had knowingly left substantial and potentially dangerous amounts of plutonium on or near the surface in parts of Maralinga.

The Americans conducted a range of top-secret 'safety trials' that investigated the one-point problem that had sparked the British Vixen B program, in a program of what they called hydronuclear experiments. Many of these trials were conducted in the early 1960s at Los Alamos. Roller Coaster was part of this broad program but was conducted (with input and participation from AWRE personnel) in Nevada and specifically examined environmental dispersal of plutonium when simulated warheads were detonated using conventional explosives. Roller Coaster was made up of four trials, held on 15 May, 25 May, 31 May and 9 June 1963. The first two were atmospheric tests, similar to Vixen B (although the cloud created in the first did not rise as high as that in Vixen B). The other two were held in bunkers. In the atmospheric tests, most of the plutonium travelled downwind as an oxide aerosol, just like it did in the Vixen B tests. The Americans were surprised by how much plutonium contaminated the steel plate in several of the tests and thoroughly studied this phenomenon. This same outcome from the Vixen B safety trials became apparent only during the 1984 ARL trip to Maralinga, even though it was there to be seen in the immediate aftermath. The actual amounts of plutonium used in each of the US tests were still classified at the time of Moroney's analysis, but he was able to postulate how much was used 'with reasonable accuracy' because of his knowledge of Vixen B: 'The Roller Coaster data ... and the ... data for the five identified plumes at Taranaki agree very nicely, given the differences in the firings, and the span of cloud heights and wind speeds'. In other words, Moroney was sure that Roller Coaster and Vixen B could be usefully compared. While Roller Coaster and Vixen B yielded different official results, the American results were more robust and realistic. The Vixen B figures made no sense unless Moroney factored in a gross underestimate of actual contamination at Taranaki. When he saw what the data meant, he made his switch from stalwart AWTSC man to disillusioned critic.

The discrepancy could not be explained away by different ways of doing things, between countries and across the years. By the time the ARL scientists were on the ground at Maralinga in 1984, the measurement and analysis methods to which they had access were far superior to those available to Noah Pearce and his colleagues who prepared the AWRE report in 1968 that cleared the UK of any further responsibility at Maralinga. The ARL scientists used gamma ray detectors to measure a product of plutonium decay, americium-241, which could be extrapolated to give an exact measurement of plutonium. As Anderson explained in his story, 'The British had to rely on alpha particle emissions from plutonium which are difficult to detect'. But, as radiation scientist Peter Burns also pointed out in the story, 'They could have done radiochemistry analysis of the soil which would have given a more accurate reading of plutonium'. In addition, the British had had the results of the Roller Coaster trials, so they would have known the likely levels of plutonium.

Not long before his death in 1993, Moroney, who planned to write what he knew, prepared several aides memoire to this end. They reveal that he was always careful to ensure that the latter-day reader would understand the mindset of the times that led to the creation of the Maralinga test site. 'A major ingredient in depicting the general background is to convey some sense of the perceptions of the period, which, I suppose, allowed these nuclear tests to be conducted essentially as military operations, with the expectation on personal compliance and commitment that this implies.'

A long period of analysis of the Maralinga site had ensued after the 1984 ARL visit. The ARL scientists discovered a major discrepancy between the levels of contamination claimed in the Pearce Report and what they found on the ground, sparking years of investigation that culminated with Moroney's detailed examination of Roller Coaster records. The americium-241 levels obtained by the ARL scientists showed the Pearce Report data about plutonium levels were incorrect. Moroney's analysis of Roller Coaster later confirmed that they were in error by a factor of 10. While the Pearce Report claimed that about 20 of the 22 kilograms of plutonium had been buried, an estimated 20 kilograms was later found to be scattered around the site.

MARTAC, set up in 1993 to oversee clean-up of contamination at the site, extended and corroborated the readings from the 1984 ARL visit and Moroney's analysis. The MARTAC team found the 'contamination of the lands consisted of fine particulate of plutonium and fragments of paraffin wax, lead, light alloys and plastic with plutonium plated on them'. In other words, the site was a dangerous mess. MARTAC also noted that 'it was the experimental set-up of the *Vixen B* trials that made them the principal source of lasting environmental contamination'.

Moroney heard from Williams about the prospective *New Scientist* story and was keen to co-operate. By then he had been crunching the Roller Coaster numbers for a couple of years, and sending what he had found to his Australian Government contacts, notably Pat Davoren, then manager of the Test Site Management Unit of the Department of Primary Industries and Energy. He also started preparing a briefing paper in response to the developing *New Scientist* story. He noted that the 'inadequacies and gross inconsistencies' in the set of plutonium field data used for Operation Brumby 'were not resolved between the UK and Australia at the time, even after extended re-analysis and debate'. This absence of proper contamination data for the site was a catastrophe. Moroney maintained that it was not a simple mistake. The Roller Coaster results meant that the 'AWRE knew of … the error involved', but these results 'were the subject of military security at the time and not accessible to Australia'.

These errors resulted in considerable confusion and misinformation about plutonium contamination at Maralinga. The Roller Coaster trials gave Moroney the ammunition he needed to support the Australian Government in rejecting the 1968 agreement. In a letter to Davoren on 28 November 1991, Moroney gave his 'fast first pass through the Roller Coaster information'. Before commencing RADSUR, the 1966 radiological survey, and Brumby, the 1968 clean-up, he concluded, the AWRE certainly knew that

» α-survey [alpha-survey] monitoring of Pu [plutonium] fallout on soil can be expected to underestimate the Pu surface density by an order of magnitude, even when the survey is made in the day or so immediately following deposition; and

» less than 20% of the Pu used in the Vixen B trials can be expected to have remained in the debris in the locality of the firing pads.

It follows from this that AWRE also knew that:

» all of the post-firing α-survey data from Vixen B trials were low by at least a factor of ten;

» the areas of Pu contamination at Taranaki to be cleaned up in BRUMBY were greater, by an order of magnitude, than as indicated by the results of the post-firing α-surveys; and

» the burial pits at Taranaki contained no more than 15% to 20% of the Pu used in the 12 firings.

Those few lines written to Davoren distilled what Moroney had deduced from the Roller Coaster pages. His analysis discovered the key to understanding what damage Vixen B had done. He subsequently supplied this information to Ian Anderson, and it provided the backbone for his article. Moroney's analysis was later confirmed by MARTAC.

In his letter to Davoren, Moroney observed, 'I know that the Pu survey work in RADSUR & BRUMBY had its problems, but I still find it galling that it was so bad that it couldn't even pick-up an error of such huge dimensions'. Moroney signed off his letter with the handwritten words 'good luck'.

The analysis showed that the Australian authorities had not possessed all the relevant facts before signing the 1968 agreement. The accumulated data from the 1984 expedition and from Roller Coaster data produced an irrefutable case for the British to help fund a major site clean-up. Simon Crean and Gareth Evans were armed with these facts as they prepared to negotiate with the British for compensation. Behind-the-scenes wrangling over this issue had been going on for several years, with expert input from Moroney. The Australian Government alerted the British Government that it had the Roller Coaster data in December 1991.

The moves to deal with the contamination began in the wake of the Royal Commission when a group known as the Technical Assessment Group (TAG), made up of British, American and Australian scientists and technicians, undertook extensive tests across a range of issues. TAG did not itself include formal Indigenous representation, as suggested by the Royal Commission, although Maralinga Tjarutja people were represented on the Maralinga Consultative Group, a broader forum to discuss all matters concerning the test sites.

TAG carried out six studies from 1986 to 1990, including inhalation studies, flora and fauna surveys and a detailed anthropological study. The group devised 27 clean-up options, preferring one that involved, in part, immobilising the waste in the Taranaki burial pits using an innovative technique known as *in situ* vitrification. Anderson reported on the outcome of the TAG investigations, in a *New Scientist* story on 17 November 1990. He highlighted TAG findings that suggested 'Aboriginal children would receive doses of radiation more than 300 times the accepted limit if they were to live in the most highly contaminated regions of the former British nuclear test site at Maralinga in South Australia'. Anderson's 1990 story also reported that the Australian Government was seeking compensation from the UK to help cover the cost of the clean-up.

It took three years. During those frustrating years – 1990 to 1993 – Moroney analysed the Roller Coaster data in detail, comparing them with the Pearce Report, the data generated by his ARL colleagues in 1984 and material that was emerging from TAG. In the process Moroney's attitude to the British tests changed. This shift was profound and painful. He deeply resented the British lies and felt that he, personally, had been misled through years of loyal service. He became a bitter crusader for accountability and nearly single-handedly pointed out all the ways the British test authorities had been deceitful. In a series of reports, analyses and memoranda he summarised the issue in a way that Australia's legal and scientific representatives were able to use.

Geoff Williams, who worked with Moroney for years, said the 1984 findings and the subsequent Roller Coaster revelations had been 'a great eye-opener' to Moroney and confirmed that he was angry. Ultimately he felt the British must have known and had 'entirely hoodwinked him and his committee … John felt very let down by the British because he felt that it was a relationship of trust. He trusted the British, he felt the British trusted him, and there was this great breach of trust where they had really done things out at Maralinga that he wasn't aware of'.

After Anderson, acting upon his chance conversation with Williams, decided to pursue the story, he interviewed Williams and Peter Burns together. In a recording of the interview, Anderson can be heard quietly but determinedly directing the two radiation scientists to tell their detailed and damning story. The scientists were calm but displeased. They were frustrated that their work back in 1984, the following Royal Commission and the extensive, exhaustive work by TAG all seemed to be coming to nothing. Williams later observed in 2004, 'We all felt originally that the British were going to get away scot-free again'. Certainly, in 1993, it was by no means certain that the truth would out.

Anderson portrayed the interviews as 'a cat and mouse game' because he was denied direct access to the declassified Roller Coaster documents. He had gone to his first meeting at ARL expecting to be shown the documents, but he never saw them. He said in his application for the Michael Daley awards, 'After that setback, it was a matter of piecing together the main thrust of the documents from what Moroney's colleagues at ARL felt they could say'. He said, 'My sources wouldn't always tell me what was right, but would indicate when something I put to them was wrong'. The existing tape recording does not bear this statement out, however. The interview sounds more open – Williams and Burns were talkative, informative and expansive. Anderson was likely unhappy he could not see the actual Roller Coaster documents himself, and it possibly coloured his perception. He believed that a 'senior bureaucrat' in Canberra had prevented him from getting the documents.

A cat-and-mouse game is certainly evident in a taped telephone interview between Anderson and an unidentified contact, however. The source was obviously a ministerial adviser who could have been any of several advising Crean at the time. Anderson (IA) was trying to find out from the unknown interviewee (UI) how much compensation was being sought from the UK.

IA: So are we going for this $101 million?

UI: Thereabouts, yes.

IA: But how much are we asking them for?

UI: A substantial contribution. You would have seen the newspaper reports about that.

IA: There was something in the *Canberra Times* about $60 million I think it was.

UI: That's inaccurate. In fact most of the newspaper reports are inaccurate – most have guessed at what a 'substantial proportion' is.

IA: Okay, well what is it then?

UI: Well, that's the Australian Government's position and up to negotiation between the two governments. We've told them what we are expecting – all we've been saying is that we are expecting a substantial ...

IA: So the idea of it being 50 per cent is not necessarily correct?

UI: No, in fact that is quite incorrect. It is certainly a lot more than that.

IA: A lot more than 50 per cent?

UI: Yeah. Simon [Crean] spoke to a number of reporters who were out at Maralinga and they reported figures of anywhere from 50 per cent to three-quarters. I would suggest that three-quarters was a far closer figure.

In his story, Anderson noted a likely compensation payment from Britain of £33 million, a figure that was roughly three-quarters of the estimated total cost. The final amount provided by the UK Government was somewhat less – £20 million, or about A$45 million – just under half of the actual cost.

Moroney became ill around March 1993, and Anderson never met him, though they did speak by phone several times. What at first seemed to be a severe case of pneumonia turned out to be multiple myeloma. Moroney died within days of Anderson's article coming out, aged only 63. He saw a draft of the story that Anderson sent to ARL for checking and clearance. The annotated draft of this document – constituting a revealing three-way conversation between Anderson, Moroney and the deputy news editor for *New Scientist* in London, Jeremy Webb – showed Moroney, in diplomatic language, savaging an early version of the story. He suggested that large swathes be removed, including the emphasis on the Pearce Report, which by now he had personally dismissed: 'I don't completely understand why we spend so much time debunking Pearce. Is this because the Brits still think this is the definitive study on Maralinga?'

In fact, the document showed that Moroney suggested that nearly 50 per cent of the draft article be cut or greatly altered. Anderson disregarded most of these suggestions and changed only things that Moroney had shown were definitely wrong or skewed. One of Moroney's colleagues was quoted in the article, and Moroney expressed some affront to him for allegedly stealing the limelight. Apart from this, though, he appeared happy with Anderson's work, as were the other scientists who informed the story.

Anderson did not mention Moroney by name in the story, despite extensive dealings with him, which is a bit of a puzzle. Anderson later wrote an account of the article in which he said that 'the story was confirmed by Moroney over the phone, although he did not want his name mentioned', and so he used other names as sources. The gusto with which Moroney approached his 'edit' of the draft suggests he was not timid or media-shy. He had been shocked by the betrayal by the British that he had played a crucial role in revealing. Yet he did not ask for his name to be added when he looked over Anderson's draft, despite suggesting major changes, and telling his colleagues that the credit for the uncovering should be his. The contradiction may be explained by Moroney's longstanding career in secret nuclear business, cut across by his anger at the British. Was he defensive about being seen as gullible now the lies were revealed? The truth is unclear.

Moroney was a complex character who straddled two distinct eras in Australian relations with Britain. He may not have given Anderson a definite signal about what credit he wanted, but he undoubtedly provided the deep background that gave Anderson's story its authority. He wasn't completely cut out either. Anderson mentioned Moroney in his ABC Radio National *Science Show* broadcast, and in his application for the Michael Daley awards for science journalism. And Moroney was due to appear on ABC TV's *7.30 Report* when the story appeared, but this was cancelled because of his ill health.

As mentioned, Anderson's article was published at a crucial time in the Australian ministerial-level negotiations with the British

Government on Maralinga compensation. Simon Crean's staff faxed the article to him in Europe, and it appears to have had an impact, though how much is difficult to measure. Anderson, a modest man, confirmed later that year that the article 'added to the moral pressure that parliamentarians and others were bringing to bear on the British government to acknowledge its responsibilities and pay up'. But Anderson said that while it might look otherwise, the timing was not a deliberate strategy: 'There was no collusion and the article was never mentioned in the negotiations'. The story came out just five days before the bilateral talks.

For Geoff Williams the article pulled together, for the first time, many of the threads of the Maralinga story and, in Anderson's words, was 'the first public airing of the betrayal by the British'. Tim Thwaites, in his *New Scientist* obituary for Anderson, said he 'put pressure on the UK Government to make a significant commitment to cleaning up the nuclear test site at Maralinga'. In an obituary in the *Guardian*, Philip Jones claimed that his 'evidence, and the media attention engendered by the material in such a prestigious science journal, played a crucial role in the successful conclusion of the talks'. When 'Britain's dirty deeds' first came out, the *Guardian* was one of several UK newspapers that cited it in stories on the Maralinga negotiations.

Maralinga featured prominently in the Australian media once the article came out, leading to a marked revival of interest in the aftermath of the British tests. This matched the earlier heated coverage when the Royal Commission was taking evidence in 1984 and when it reported in 1985. Anderson personally promoted the story in various ways. On the *Science Show* he gave a radio-friendly summary about the significance of the article, saying, 'Australia, represented by foreign minister Gareth Evans and energy minister Simon Crean, will present a strong and compelling case to Whitehall'. He also credited John Moroney and his lengthy involvement with Maralinga. He wrapped up his *Science Show* talk with a flourish:

If Australia is right, Britain misled a true and trusted ally and that ally is now paying for that trust. In monetary terms, Australia itself is facing large payouts as veterans of the British atomic tests at Maralinga press their claims in court. But will Britain pay its share for another clean up? Will it pay compensation to the Aborigines? Recent statements in the British Parliament suggest that it will not. It will stick to its belief that its obligations have been met. It's just not cricket.

Australian metropolitan newspapers picked up the allegation of British deceit and the abundance of abandoned plutonium at Maralinga. A feature in the *Sydney Morning Herald* on 10 June 1993, prompted by a preview of Anderson's story, revealed the fascinating fact that Dr Mike Costello, who worked for TAG, had probably helped create the Maralinga plutonium that was now causing so much controversy. The story reported that Costello had been a chemical engineer who had worked on plutonium for the UK Atomic Energy Authority in the late 1950s. The next day, a news article in *The Age* said the negotiations with Britain 'could have been strengthened by new evidence' in Anderson's story and quoted the South Australian minister for Aboriginal Affairs Kym Mayes saying, 'the British Government could not ignore the magazine's allegations'.

Anderson's article was not the only source of pressure on the UK Government, however. In 1993, a delegation of Aborigines from the Maralinga lands (including the prominent Indigenous activist Archie Barton) had arrived bearing sand from the region – not actually contaminated sand – which they placed on the steps of the British Houses of Parliament. British parliamentarians, notably the outspoken minister of state for the Armed Forces Archie Hamilton, had been asserting that the 1967 clean-up had been effective. Their message was undermined, though, when the government called in people wearing full contamination suits to remove the sand from the steps. As Peter Burns remarked, 'They had said it was all right to live in this sand 24 hours a day, 365 days a year, camp in it, eat in it, hunt in it. But as soon as they put a few kilos on the steps they got guys in decontamination suits. Talk about a PR disaster'.

Pat Davoren, at the time responsible for co-ordinating the development and presentation of Australia's Maralinga case to Britain, confirmed the impact of Aboriginal delegations outside parliament at various times during the dispute: 'I got the impression from British Ministry of Defence officials that these visits did have some effect (they wished they would stop!)'.

Anderson's story was timely, terse and filled with cross-checked data. Its sub-heading summed it up: 'Fresh evidence suggests that Britain knew in the 1960s that radioactivity at its former nuclear test site in Australia was worse than first thought. But it did not tell the Australians'. The two-page news feature had a generic picture of an A-bomb mushroom cloud, a graphic map of the radiation plumes that emanated from the Taranaki test site and a picture of two unidentified scientists collecting samples during the 1984 survey at Maralinga. The science was deftly woven into the politics and the history:

> Burns and his colleagues now believe that contamination at
> Maralinga is much worse than Britain has admitted. They say
> 21 pits, which were dug to hold radioactive waste, contain
> far less plutonium than Britain maintains. The remaining
> plutonium – ten times more than Britain has acknowledged –
> was spread over the land. The Australians will say that if
> they had known the full extent of the pollution, they would
> never have signed the agreement releasing Britain from its
> responsibilities over the cleanup.

Jeremy Webb, who edited the article, having quite a bit of input, as was usual for *New Scientist* editors, remembered its bombshell effect. He noted that publishing just before the bilateral meetings was critical in creating a storm: 'The injustice was blatant and the story was widely covered. Obviously the British government would have preferred it if the negotiations had gone on in secret. But suddenly the talks were in the media spotlight with news outlets and the public wanting to know how the wrongs would be righted'. The management of the magazine was well pleased, he said. 'There was a great sense of pride at *New Scientist* that we had helped to make a difference.'

Brian Toohey had opened a multi-faceted story to media examination back in 1978. Now Ian Anderson's contribution had provoked a new round of public and political pressure, leading to a compensation agreement with Britain. Anderson appeared in many Australian media outlets when the story was released. He told Tony Delroy's audience on ABC Radio National's late evening show on 10 June, 'This story as you know has been bubbling away for quite some time. Little bits and pieces have come out. What we have got here I think is just a pulling of it together'. He also discussed how the recently declassified documents had helped to bring out the truth.

Asked by Delroy about the forthcoming intergovernmental meeting in London, Anderson continued his favoured cricketing theme as he pondered the battle ahead: 'A very tough fight, yes. Ironically the Australians will be doing battle with the Brits at Lords the same time won't they?' He pointed to a debate on 1 April in the UK parliament, discussed in his story, which had made clear that the British believed the 1968 agreement and the Brumby clean-up had fulfilled their responsibilities. The British had also denied that they were responsible for compensation to the Aborigines.

In the debate, Archie Hamilton had stood up in the House of Commons and maintained that Britain should not and would not pay. He had quoted from the 1968 agreement signed by the governments of Australia and the UK:

The United Kingdom government have completed decontamination and debris clearance at the Atomic Weapons

> Proving Ground Maralinga to the satisfaction of the Australian
> government ... With effect from 21 December 1967, the
> United Kingdom government are released from all liabilities and
> responsibilities under Memorandum of Arrangements save that
> the United Kingdom will continue to indemnify the Australian
> government in accordance with Clause 11 of Memorandum in
> respect of claims for which the cause of action took place after
> 7 March 1956 and before 21 December 1967.

(Clause 11 specifically guaranteed to provide compensation for claims of death, injury or property damage sought by British government employees only, something that the Australian Government had agreed to despite an initial weak protest.) Hamilton had also mentioned how Britain had already repatriated half a kilogram of plutonium from the site in 1979, and how this was followed by an 'exchange of notes' that stated there was 'no question of the United Kingdom having any further responsibility to repatriate waste'.

In the *New Scientist* story, Anderson had paraphrased Hamilton, who had equated dose levels at Maralinga 'to those in Cornwall from naturally occurring radon gas'. He had also given Geoff Williams' rebuttal, which had described Hamilton as mischievous: 'It is not acceptable internationally to compare levels of man-made radioactivity with those of a naturally occurring radionuclide ... Doses in Cornwall could reach 8 millisieverts a year. But, according to the TAG, because of the Aboriginal lifestyle, a child living near Taranaki could inhale more than 460 millisieverts a year'. Young children were at the greatest risk, because they were closer to the dusty ground and had smaller body mass.

Anderson was asked several times in different interviews to speculate on what the British knew and when they knew it. He tried to be balanced and fair. When asked by one host if it was proved that the British had knowingly lied, Anderson replied:

Well that's a very, very good question. Was it deceit or not? You
have to go back to the time ... the world was different 30 or
40 years ago. These were cloak and dagger days and it has been
suggested to me particularly by a person who was involved a
lot on the Australian side at the time that the British Atomic
Weapons Research Establishment, or parts of the Establishment,
may not have been talking to each other. So whether it was
deceit and deliberate is another matter – I think the crucial thing
from Australia's point of view is that it happened, and therefore
40 years on Australia believes it has a moral right for the British
to participate again in a cleanup.

Almost certainly Anderson was referring here to Moroney, who had
guided much of his understanding of the issue. The host continued
with the theme of deceit, commenting that the key question was
British culpability 'and the extent to which Australian officials may
have been part of that conspiracy of silence'. Anderson said:

Whether there was any Australian duplicity in it is another
interesting point ... I guess one of the questions that comes up
is: why didn't Australia do a more thorough job itself at the time
and find out what was going on back then? Of course I get back
to the point that this was a long time ago. I think that the British
position was probably to a large extent taken and not questioned.
We were much, much closer to the British in those days – in fact
it was suggested in the parliament in London the other day that a
lot of this, as far as the arrangement to do the testing, was stitched
up in a telephone call between Robert Menzies and Clement
Attlee, who was the British PM at the time. I doubt very much
whether telephone calls these days would come to such deals.

In an interview on South Australian radio station 5CK, Simon
Royal raised the slightly qualified *New Scientist* editorial support.
The magazine's editorial had suggested that 'even if Australia has

right on its side, it is too much to expect that Britain should immediately offer to pay for part of the clean-up. The sums of money are
not massive by government standards but they are far too big for the
Treasury to part with lightly'. The rest of the piece had supported
the story. Royal asked Anderson how he felt about the suggestion
that Britain shouldn't immediately offer to pay up. Anderson replied
that the editorial had been suggesting Britain come clean first and
'then, probably, pay up'. Royal asked the question again and Anderson gave his own view:

> I think that, in my own personal opinion, the British should pay
> up, that it is quite clear that the cleanup that was done, Operation
> Brumby in 1967 and the report that was done into it by Pearce
> in 1968, it wasn't correct. For various reasons the cleanup
> was not done properly. Now we have the technology to do it
> properly, and Australia I don't think has been unreasonable – it
> was presented with a range of options from about $13 million
> to about $600 million to clean the place up and it has chosen, if
> you look at the document, bits and pieces of this and that and
> come up with $101 million. And that to me seems a reasonable
> amount and really by today's international standards it is not a
> huge amount.

Royal was also interested in the role the article might play in the
ministerial discussions about to get underway in London, asking if
the article added more fuel to the fire. Anderson replied:

> I should think so, yes. *New Scientist* is quite widely read in the
> UK – it goes to Whitehall in other words. The point is, why we
> concluded that it was going to be a heated meeting is that in
> all the public statements that have come out recently, especially
> in the parliament over there, it's quite clear that unless there's
> something going on behind the scenes, but at least publicly they
> do not intend to pay up.

The British Government faced other forms of pressure too. A documentary prepared by the BBC, with Australian Government assistance, entitled *Secrets in the Sands*, was broadcast in Britain on 28 October 1991. It revealed the human and environmental toll of the British tests and was screened just before Crean met Lord Arran, then undersecretary of state for Defence and the Armed Forces, to present a case based on the TAG report and early interpretations of the Roller Coaster data. The material in the *New Scientist* story had been in official British hands since 1991, having been presented by a senior Australian Nuclear Science and Technology Organisation scientist, Des Davy (then general manager scientific for the organisation, and also convenor of TAG), during an official meeting in December 1991. But the information was not made public. The public and private pressure on the British Government was mounting at this time, far more than it had done during the Royal Commission some years earlier. Anderson's story seemed to cap off the moral case that Australia had been making for compensation. Less than a week after the *New Scientist* story appeared, compensation was finally promised.

'Britain's dirty deeds at Maralinga' was an important piece of scientific investigative journalism. It contributed (either directly or indirectly) to a political solution to a longstanding national problem. The story resonated beyond the *New Scientist* readership, becoming a high-profile mainstream media story in Australia and adding to the body of investigative journalism that finally illuminated Maralinga. The story also provided conclusive proof that the old way of reporting on the British nuclear tests in Australia was gone. Accepting official information, explanations and undertakings was no longer sufficient. The journalists covering the tests and their aftermath now were watchdogs and, true to the metaphor, were dogged in seeking the truth. Anderson's story was a pivotal moment in the uncovering of Maralinga, marking at last the full transition from opacity to transparency. In this sense it was the culmination of a process that had begun 15 years earlier. Although the Maralinga lands may not have been completely remediated by the compensation deal that was finally struck a few days after the story was published, more was done than if the issue had languished without such intense public scrutiny. The outrage of a wronged servant of the British nuclear tests, John Moroney, found the right outlet. Much to his frustration, terminal illness prevented Moroney from playing the central role he believed his involvement warranted. He did not live to see the outcome of his painstaking Roller Coaster analysis, either, but one must imagine he would have been well pleased.

1 2

The remains
of Maralinga

*The little bridge they crossed on the oleander-lined path leading
from the airfield to the terminal was called the Bridge of Sighs. Last
rites – a sigh of trepidation by those arriving; a sigh of relief by those
departing – were often performed on that spot.*

John Keane, 'Maralinga's afterlife', 2003.

*The whole story, when one looks at it in detail, is rather sordid and
the major villain in that sordid story is without doubt the then Prime
Minister in the early 1950s, the lickspittle Empire loyalist who
regarded Australia as a colonial vassal of the British crown. I refer
of course to Sir Robert Menzies, the twentieth century satrap who
invited the British to pollute Australia with nuclear fallout:
the pseudo patriot who cravenly surrendered Australian sovereignty
to a declining imperial power.*

Senator Peter Walsh, minister for Mines and Energy,
Australian Senate, 1984.

Thomas Tooke has been with the RAAF since 1943 and has seen
service in Korea and Japan. Now he is sent to Maralinga as a
despatch corporal, having been kitted out at the Edinburgh RAAF
base north of Adelaide with an extra pair of drab pants, a shirt and

a pair of boots. He is bemused that when he arrives at Maralinga in 1956 he is issued with an army uniform as well, without explanation. He has a bigger shock when he arrives at Camp 43, not far from the forward area. The bulldust is like nothing he has ever seen. It's as fine as talcum powder and gets into everything. He and his comrades find that the bulldust conceals a layer of hard limestone when they try to drive tent pegs into the ground. They have to get jackhammers to make holes in the limestone to raise their two-man tents. They get some gravel from the Watson railway siding to try to damp down the swirling, annoying dust.

The men eat in the marquee and the food's okay. The blowflies, though, are terrible. Unfortunately, they 'blow' the food by laying maggots in it, and bad things happen as a result. He and everyone else he knows have bouts of terrible diarrhoea. He drops from 86 kilograms to 70. That is not the only discomfort. Any metal at the camp is so hot that you can't touch it, as temperatures soar above 38 degrees Celsius. The open showers have a drum holding several gallons of muddy, salty water held aloft by a hook on the side of a tent. If it's windy when you take a shower, you have to follow the droplets of water around before they are carried away. There is a coconut oil soap called Seagull Soap, which can lather even in the hard water. Every fortnight on pay day the men get one cake of soap and two razor blades. Tooke hears a rumour about an attempted lynching of a civilian, one of the construction crew, caught cheating at cards. A South Australian policeman apparently stepped in and stopped it. The desert conditions make everyone a bit crazy. Lennie Beadell swings by every so often, with his Land Rover packed to the gunwales, on his way to an even more remote location.

After a while, Tooke moves from the tented camp to Maralinga village. It is a bit more luxurious, but the aluminium sheeting on the roof constantly lifts up and flaps around, requiring endless running repairs. Still, most of the cooks in the village are navy men, and the food is excellent. There are movies six times a week, although sometimes the same film is shown two nights in a row. As the day approaches for the first Maralinga atomic test, the village fills up. Boffins from Britain start to arrive, and observers from New Zealand, America, Canada and other places. Even the observers are pressed into service. No man is left idle; they all get to work on myriad construction tasks. Tooke drives his 10-tonne crane to the forward area. He sees a working group and asks if they know where a colleague is. A cultured voice replies, 'What regiment is he with?' Tooke drives off laughing.

One Tree at Maralinga, 27 September 1956. Finally the winds have died down and the countdown begins. Tooke is less than a mile away from the forward area when the awe-inspiring explosion takes place, and eight hours later he enters the forward area with his crane. There are people everywhere, carrying out lots of different tasks. Tooke must recover vehicles deliberately placed there. The health physics people have given him bootees to wear, the only items of protective clothing he is ever given. His other garb is his RAAF overalls. He never receives a film badge or dosimeter. Most of the people sent to recover vehicles from the forward area are RAAF personnel. Some of the vehicles have been tipped over onto their sides. There are tanks, Land Rovers, Commer vans, artillery, anti-aircraft guns, Humber Scout cars and even Swift aircraft. After he comes back from work that day, a Geiger counter is run over him and it clicks slowly. The new, expensive, permanent nuclear test site in the South Australian desert is now fully functional. Tooke has a lot of work to do.

The atomic age arrived when the US dropped nuclear bombs on the Japanese cities of Hiroshima and Nagasaki in August 1945, heralding an era that seemed to be even more dangerous than the war just ending. When nuclear scientist spies wreaked havoc on postwar security, the British, sidelined by the Americans, were on their own. They embarked on their pigheaded, quixotic, ill-advised, careless but still rather remarkable quest to match the Americans and the

Soviets in the nuclear club. In fact, they were not entirely alone, as they co-opted the Australian Government for the task. At all times, the relationship between the British and the Australians was unequal. The British were the masters, and the Australians were the servants. The Australians obediently provided the site and considerable finan-cial and military resources, as well as staunch political backing. The atomic test authorities made the decisions and relayed them, often with inadequate technical detail, to the Australians. When the tests were over and the British were gone, the picture of what they had left behind was alarmingly incomplete. The Pearce Report was no help. What exactly the tests had wrought remained hidden to suc-cessive Australian governments and the Australian people for far too many years.

But, one by one, the jigsaw pieces fell into place and the Maralinga story started to take shape. The British authorities have still not given up all the missing pieces. No-one outside a small circle knows why the UK Ministry of Defence still retains some files relating to Vixen B and other issues (including information recently released by WikiLeaks). Whether those files will ever be released is currently unknown. Given Vixen B's impact on Australian territory, the ongoing refusal to release all the information that relates to those experiments could reasonably be called outrageous. This saga tells us, though, that the British authorities charged with testing the nuclear deterrent did not factor in Australian feelings. The truth is unpalatable but must be faced: Australia in the 1950s and early 1960s was essentially an atomic banana republic, useful only for its resources, especially uranium and land.

Australia tried without the slightest success to have some status in the arrangements. It was not to be. When the world turned and Australia no longer had anything vital to offer, the British left with-out properly cleaning up their mess. Tellingly, the legendary offi-cial British historian Margaret Gowing rarely mentioned Australia throughout her magisterial accounts of the British nuclear enter-prise. Australians were not a partner in any sense of the word, just lackeys and useful idiots for the most part. All the historical circumstances that converged after World War II made this inevitable, and it should not perhaps be surprising that this was the reality.

Other realities have to be faced too. Australia, not Britain, was left with severely contaminated territory requiring remediation that took several years to complete. Money was finally squeezed out of a resistant UK, but it covered less than half the cost of the most recent clean-up. And the controversies continue. The 1990s remediation plan was devised using guidelines provided by the IAEA and the International Commission on Radiological Protection. It was monitored by ARPANSA and agreed to by the Maralinga Tjarutja traditional owners. This plan involved securing 500 000 cubic metres of contaminated waste from the tests in 21 pits at the site. When it was completed in 2000, Australian prime minister John Howard called it the 'world's best practice clean up'. But nuclear engineer Alan Parkinson, sacked from the clean-up, and Jim Green, an anti-nuclear activist, among others, continue to criticise its inadequacies.

The clean-up project used the expensive electronic technique *in situ* vitrification, which involved placing electrodes in the pits and running an electric current of up to 4 megawatts through the buried debris, raising it to temperatures up to 2000 degrees Celsius and effectively turning it to glass. But it did not go as planned. The vitrification equipment exploded while in operation at pit 11 (the 13th 'melt' in the clean-up program) and sprayed molten material from the pit about 50 metres all around. Fortunately no workers at the site were injured. But the vitrification process was abandoned before all 21 pits were processed, and the remaining waste pits were capped with concrete. The glassy material that had been vitrified was excavated and reburied amid fears that it was too close to the surface. Recently, some media reports have claimed that the Maralinga pits are subsiding and eroding, creating fears that the contamination is not permanently secure. Nevertheless, in March 2003, minister for Science Peter McGowan triumphantly tabled the final MARTAC

Report in parliament. MARTAC hailed the clean-up as successful.

Alan Parkinson vociferously disagreed, saying that 'of the hundreds of square kilometres contaminated, only 2.1 square kilometres have been cleaned to the clean-up criterion and, of that, only 0.5 kilometres permits unlimited access. The only people who claim the [clean-up] project a success are paid by the government'. Stuart Woollett, an ARPANSA scientist involved in the clean-up, presented a different view: 'The release of the MARTAC report marks the end of the considerable work of health physicists, radiation chemists, plant operators, security personnel, surveyors, camp staff, senior public servants, seed planters – the list goes on. Operations, beginning in 1996, have rehabilitated the Maralinga lands for their return to the Maralinga-Tjarutja'. Gregg Borshmann addressed the ongoing controversy in a *Background Briefing* documentary titled 'Maralinga: the fall out continues', that aired on ABC Radio National in April 2000. Every so often, public disquiet about Maralinga still bubbles to the surface.

The Maralinga story is filled with outrages. A story that began to appear in the Australian media in September 2001 described one particularly distasteful aspect of the saga, namely the analysis of (mostly) babies' bones to detect radioactive fallout. At a meeting in Harwell on 24 May 1957 attended by Ernest Titterton, along with his confreres from the AERE and the AWRE, a variety of sampling tests was ordered, including soil, vegetation, milk and sheep bones. And babies' bones.

> As many samples as possible are to be obtained (the number available is expected to be small). The bones should be femurs. The required weight is 20–50 gm. Wet bone, subsequently ashed to provide samples of weight not less than 2 gm. The date of birth, age at death and locality of origin are to be reported.

Every cliché of the mad, obsessed scientist is entangled in that quote from the report presented to the AWTSC on 11 June 1957. This fact about the British tests ignites public sentiment like few others. Where the tests intersect with the tragedies of infant deaths, few Australians are likely to be unmoved. Many of these concerns centre firmly on Titterton. The meeting minutes recorded that 'Prof. Titterton will ensure that arrangements are made by the Australian Safety Committee for collection of all of the above samples and despatch them to the U.K. along with all relevant information, addressed to Dr. Dawson, A.W.R.E., who will pass them to A.E.R.E.'

The bones were scattered far and wide, from Australian laboratories in Adelaide and Melbourne to British ones in Aberdeen, Liverpool and London. The parents of the babies never knew. In all, bones from nearly 22 000 bodies – the majority of them babies – from both Australia and Papua New Guinea formed part of the experiments. Collection of samples in Australia was part of a bigger international program perhaps ironically titled Project Sunshine. The bones were tested for strontium-90. John Moroney from the AWTSC said in a letter to the pathologists involved (quoted in an Adelaide *Advertiser* article many years after the event), 'You may have perhaps considered it possible that the question of sampling and radiochemical assaying of bones would not be regarded kindly by the general public. Consequently, I would be grateful if … you could treat this matter as either confidential or personal'. While a rational argument can be made that testing bone for traces of radioactivity during the early days of the atomic age has valid scientific justification, no amount of reasoning is likely to reassure the families of the 22 000 babies and others whose bones were tested. Stealing the bones of babies can never be seen in purely scientific terms.

Then there are the nuclear veterans. These aggrieved men, their wives or widows and their children have never been recognised for what they endured. Service personnel who go to war have well-recognised and justifiable rights. Service personnel who stood with their backs to an atomic mushroom cloud, or who scraped the spoilt soil into pits or washed down the aeroplanes that had flown through a high atomic cloud do not enjoy the same rights. The various veterans' associations in Australia and the UK have fought against this intrinsic injustice for many years, with little success. John Keane summed it up in *The Age* in 2003:

> Five decades after entering service, the thousands of British and Australian men who have survived Maralinga (more than a quarter of them are now dead) feel hurt and humiliated. They have no medals to pass on to their grandchildren, no letters of praise or apology from Tony Blair or John Howard, no wartime veterans' privileges. What they do have are anecdotes about unusual clusters of multiple myelomas. Hip and spine deformities. Teeth that are falling out. Poor eyesight. Bleeding bowels. Post-traumatic anxiety and depression. And perhaps up to a quarter of them, according to preliminary data collected by the New Zealand government, have disabled offspring.

Despite many court cases and claims for monetary recompense, only a relatively small number of Australians – military and civilian – have been compensated for health problems alleged to have been caused at Maralinga or the other test sites. Documents associated with the Australian Participants in British Nuclear Tests (Treatment) Bill 2006 provided the following statistics:

> Since the conclusion of the British Nuclear Testing Program, at least 79 common law actions against the Commonwealth have been instituted by ex-servicemen, other former Commonwealth employees and employees of Commonwealth contractors. Many of the cases before the courts have either been discontinued or withdrawn. Four cases have been heard by the court.

In addition, compensation has been paid under an administrative scheme to a number of service personnel, Indigenous people, civilians and some families of diseased people, with an average payout of $126 561. Even fewer of the many British service personnel present at Maralinga have been compensated, owing to British laws that until 1987 limited liability for injury suffered during military service. In 1988, the British Government finally agreed to pay war pensions to service personnel who developed blood cancers after their service at Maralinga.

Yet even the McClelland Royal Commission into the British atomic tests, motivated as it was by a passionate chair who sought to assign blame to those responsible for the tests, failed to find sufficient evidence of specific harm caused. While there is much anecdotal evidence, some of which has been presented in court, proving causality is extremely difficult. As scholar Paul Brown stated, 'In a finding that continues to frustrate veterans, the Royal Commission concluded that illness, disease and abnormality cannot be unequivocally associated with radiation exposure well above the dose limit'.

The Royal Commission noted the fears that the tests had engendered in participants that stayed with them well into the future: 'Operation of the "need to know" principle and the minimal amount of information given to participants has been a factor contributing to participants' concerns and fears regarding what might have resulted from their experiences at Maralinga'.

In 2002, John Clarke QC was appointed by the prime minister John Howard to review veterans' entitlements, including those of nuclear veterans. Clarke received 160 submissions on the British nuclear tests, and a chapter of the resulting report dealt exclusively with the veterans of the tests. Clarke called the nuclear tests 'a unique, extraordinary event in Australia's history' and found that members of the armed services were exposed to dangers beyond those normally experienced in peacetime. As one of the submissions to the inquiry put it, 'Australian servicemen were provided on loan to an experimental nuclear weapon test programme under the control of another country without prior scientific examination, independent advice or assessment of the potential dangers that could occur'. The report noted that these veterans were not at that point entitled to benefits under the *Veterans' Entitlements Act 1986* because their service occurred during peacetime before 7 December 1972 and recommended that the Act be extended to enable some limited coverage. The Act was amended in 2006 to enable compensation for veterans suffering one kind of cancer, malignant neoplasia. This did not go far enough in the eyes of the veterans, who had reported many other kinds of cancers, as well as issues around fertility, genetic harm to their children and mental health. Finally, in 2010 the Rudd Labor government amended the Act again to broaden the coverage of nuclear veterans. Various nuclear veterans' associations continue to fight for recognition of the harm caused by the service, and in many cases the children of the service personnel fight too.

Tens of thousands of service personnel were based at Maralinga, flew aircraft to and from the sites or were present at the other sites during the test series. Their stories have been told in various ways over the years, beginning with a remarkable series of articles in the Adelaide *Advertiser* in 1980. Later, the academic Roger Cross joined with nuclear veteran Avon Hudson to co-write *Beyond Belief: The British Bomb Tests; Australia's Veterans Speak Out*, which revealed shocking tales of what went on. In more recent times the journalist Frank Walker has meticulously documented the stories of veterans as part of his book *Maralinga: The Chilling Exposé of our Secret Nuclear Shame and Betrayal of our Troops and Country*.

The injury inflicted upon the Indigenous people cannot be properly measured. Robbing people of their ancestral homelands, subjecting them to forced removal and, later, exposure to the plutonium-laden dust and debris is not something that can be forgiven. Maralinga Tours, a successful tourist venture wholly owned by the traditional owners, began in 2014. The venture became possible when unrestricted access to the final part of Maralinga land was finally granted to the Maralinga Tjarutja people. At that point, the traditional owners could come and go freely throughout the site, without permission from the Department of Defence. May this venture thrive and prosper, because of all the people harmed by Maralinga, the Indigenous people were the most powerless.

The toxic physical legacy of Maralinga can almost be summed up in one word: plutonium. When MARTAC reported in 2002 on the outcome of the operation to remove contamination from the area, co-funded by the British Government, it said, 'Plutonium (Pu) was almost entirely the contaminant that determined the scope of the [Maralinga rehabilitation] program. It is acknowledged as a very radiotoxic element if taken into the body, particularly by inhalation'. Plutonium-239 has significant consequences for the environment. According to radiation expert Frank Barnaby, 'To all intents and purposes, once [plutonium-239] is in the environment, it stays there permanently. Because of its radiotoxicity and long half-life the disposal of plutonium presents particularly difficult problems'. While many of the people associated with Maralinga tried to play down the risks of leaving plutonium on the open range over the years, their assurances ring hollow. This material is deadly, and even back then this was known. Why was leaving it there considered acceptable? None of the answers given over the years seems satisfactory.

Uncovering secret information is a theme throughout the saga of the British bomb. An interesting side note is provided by the whistleblower website WikiLeaks, which revealed the plans drawn up by William Penney for an atomic bomb design that became Blue Danube. It is worth tracking back a little to recall the early history of the saga. In 1947, Penney was asked by the GEN.163 Cabinet committee to head Basic High Explosive Research, which later became the AWRE. This research group was tasked with fulfilling the Attlee political ambition of turning the UK into a nuclear power. Penney drew upon his extensive knowledge of the design of nuclear weapons as part of the Manhattan Project, and particularly the Fat Man plutonium bomb that was dropped on Nagasaki, and started to sketch out a design. This remarkable document, titled simply 'Plutonium Weapon – General Description', included a sketch of how the weapon might be constructed, although without great technical or scientific detail. The report was declassified and made publicly available some years ago, but the UK Ministry of Supply suddenly withdrew it from public view in 2002 (along with many other files relating to the British nuclear tests in Australia retained by other government entities, particularly the Ministry of Defence). Wiki-Leaks, however, published the full report, including the drawing, on its website in March 2008, arguing that, even though it had been withdrawn from the UK Public Records Office, the file was in the public domain since no attempt had been made by the UK Government to track down the many copies circulating since it was first made public. The government took an interest in clamping down on its distribution only when WikiLeaks published it.

There followed a bizarre and archetypically British correspondence between WikiLeaks and the head of the UK Foreign and Commonwealth Office Counter Proliferation Department, Regional Issues. Blue Danube had been superseded decades earlier and was no longer part of the British nuclear armoury. The office head said, 'I have had an initial assessment from our experts. They are extremely concerned by the drawing you have posted on your website and assess it is of serious proliferation concern, and possibly terrorism concern'. WikiLeaks 'did not find [the concerns] credible' and refused to remove the document. It remains there at the time of writing.

However, the emails between Jay Lim at WikiLeaks and Isabelle McRae at the Foreign and Commonwealth Office showed something of a half-hearted attempt at information control by the British authorities, perhaps in contrast to the era of the nuclear test series in Australia. McRae responded to Lim's initial refusal to remove the material by saying:

> I will talk to our experts here and do my best to work up a
> detailed explanation for you (though some of the explanation
> may be classified!). I am glad to read that you have at least
> checked this with a number of nuclear physicists before putting
> it on your website. I would just add that I don't see that the

information furthers your aims – i.e. reduced corruption, better government and stronger democracies. Therefore, I would be very grateful if you could remove the information while I work up a detailed explanation for you. I will try to do this as quickly as possible – I am away over Easter but if you could give me until 2 April, I'll send you something then.

Apparently bomb design–seeking terrorists observe Easter breaks too. Lim replied, 'After consultations it strikes us as extraordinary that the FCO [Foreign and Commonwealth Office] claims the WikiLeaks documents are a proliferation issue worthy of censorship, but, apparently, not worthy of assigning a staff member to address the issue during its Easter break'. WikiLeaks refused to budge, saying that 'the documents are a substantial piece of world history and have been released, then censored. Implicit in our core mission is preventing censorship of such documents'. The issue of British nuclear weaponry remained controversial, long after the British gave up testing on Australian soil.

Britain's nuclear program evolved rapidly from those Australian beginnings. In 1963, Britain purchased Polaris missiles from the US and added its own nuclear warheads, an arrangement that flowed directly from the resumption of nuclear weapons co-operation in the late 1950s. These submarine-based weapons became the basis for the country's nuclear deterrent between 1968 and 1996. The new co-operative phase did not last, since Harold Macmillan's successor, Harold Wilson, was less inclined to pursue further nuclear weapons development with the US. Polaris was bolstered by an improved design known as Chevaline, which had been tested in Nevada in the 1970s, and was later superseded by the Underwater Long Range Missile System, better known as Trident, in the early 1990s, all submarine-based weapons. The future of the aging Trident weapon is currently the centre of ongoing political tensions.

Hardship often brings out the best of creativity in people. Maralinga has sparked beautiful art and beautiful music. A travelling exhibition titled *Black Mist Burnt Country*, with plans to run for two years from September 2016, honours the output of many artists moved by the legacy of Maralinga. A long-term creative project called Nuclear Futures, which began in Australia but has grown to encompass six countries in all, 'supports artists working with atomic survivor communities, to bear witness to the legacies of the atomic age through creative arts'. A piano and violin piece titled *Maralinga*, composed by Matthew Hindson, was performed by the Australian orchestra Ruthless Jabiru in London in October 2013, conducted by Kelly Lovelady, in a program titled *Maralinga Lament*. Novels have been written about Maralinga, notably *Maralinga* by Judy Nunn and *Maralinga My Love* by Dorothy Johnston. A theatre performance produced by arts company Big hArt titled 'Ngapartji Ngapartji' premiered at the Melbourne International Arts Festival in 2004, partly in Pitjantjatjara language. In August 2006, Paul Brown's 'Maralinga', a verbatim play developed with the Maralinga Research Group based on the experiences of nuclear veterans, premiered on the Central Coast of New South Wales, directed by Wesley Enoch.

Australia was no doubt exploited by its former colonial master, but the country willingly allowed it to happen and even paid to be involved by setting up the Maralinga range and providing various kinds of personnel and logistic support throughout the test series. Why? What does Maralinga tell us about our nationhood? From this distance, the events of the test era speak of a somewhat immature democracy, anxious to please its motherland despite the high cost. Most of the decisions about the atomic tests taken by the Australian Government were not discussed and debated in public. The secrecy put in place at the atomic test sites, shored up by the imposition of information controls such as D-notices that deliberately fostered media self-censorship, enabled experiments of unprecedented risk to be conducted without public consent and their aftermath to be left unaddressed for many years. On the dusty and expansive desert test range, experiments on the destructive capacities of the atom proceeded without complete safeguards, including the safeguards afforded by public scrutiny and accountability.

Could harm of this kind happen again? The answer must be yes. Without independent scrutiny of their activities, governments are capable of anything. In more recent times, the Edward Snowdon revelations about US and UK government surveillance of citizens and the leaders of other countries gave the world a glimpse into a covert world of government activity that had, until that moment, been invisible to the majority of people. Snowdon 'revealed to Americans a history they did not know they had', as one of the journalists who received the leaked material said; the nuclear veterans, Indigenous people, journalists and politicians who blew the whistle on the British nuclear tests did the same in Australia.

The hazards posed by the tests were significant and continued for many years. However, these intrinsically dangerous experiments were not available for public assessment largely because the media, in line with official British and Australian government policy, did not report them to the public. The fact that their dangers and damage were not part of Australian public conversation had dire ramifications. A deadly substance was scattered across the Maralinga lands, and an equally toxic legacy of cover-up and deceit was left behind. To this day, we do not know the full extent of the human toll. Australia fulfilled the role its government had volunteered it for 11 years earlier, but the cost was immense. If there is a word that speaks not only of thunder but also of government secrecy, nuclear colonialism, reckless national pride, bigotry towards Indigenous peoples, nuclear era scientific arrogance, human folly and the resilience of victims, surely that word is Maralinga.

APPENDIX

British atomic tests in Australia

MAJOR TRIALS

Operation Hurricane

Monte Bello Islands, Western Australia
3 October 1952

Operation Totem

Emu Field, South Australia
Totem 1: 15 October 1953
Totem 2: 27 October 1953

Operation Mosaic

Monte Bello Islands, Western Australia
Mosaic G1: 16 May 1956
Mosaic G2: 19 June 1956

Operation Buffalo

Maralinga, South Australia
Buffalo 1 (One Tree): 27 September 1956
Buffalo 2 (Marcoo): 4 October 1956
Buffalo 3 (Kite): 11 October 1956
Buffalo 4 (Breakaway): 22 October 1956

Operation Antler

Maralinga, South Australia
Antler 1 (Tadje): 14 September 1957
Antler 2 (Biak): 25 September 1957
Antler 3 (Taranaki): 9 October 1957

MINOR TRIALS

Kittens

Emu Field and Maralinga, South Australia
Emu Field: September–October 1953
Maralinga (Naya): May–June 1955

(Naya): March 1956
(Naya): March–July 1957
(Naya): March–July 1959
(Naya): May 1961

Tims

Maralinga, South Australia
(Naya): July 1955
(Kuli/Naya): March–July 1957
(Kuli): September–November 1957
(Kuli): April–June 1958
(Kuli): September–November 1958
(Kuli): May–November 1959
(Kuli): April–October 1960
(Kuli/Naya): August 1961
(Kuli): March–April 1963

Rats

Maralinga, South Australia
(Naya): April–June 1958
(Naya): September–November 1958
(Dobo): March–July 1959
(Naya/Dobo): September 1960

Vixen A

Maralinga (Wewak), South Australia
June–August 1959
May–August 1960
March–April 1961

Vixen B

Maralinga (Taranaki), South Australia
September–October 1960
April–May 1961
March–April 1963

Glossary

Alpha particles   Positively charged particles containing two protons and two neutrons that are emitted by certain radioisotopes, particularly those with a high atomic number.

Alpha radiation   Radiation caused by alpha particles. Alpha radiation has very little penetrating power but may present a serious hazard if alpha particles are inhaled or ingested.

Atom   The smallest particle of an element that retains the characteristics of that element. It is made up of a nucleus and a cloud of surrounding electrons.

Atomic number   The number and position of an element in the Periodic Table, equating to the number of protons in the nucleus.

Becquerel   The international standard unit of radioactivity, defined as one radioactive disintegration per second.

Beta radiation   Radiation caused by beta particles. Some radioactive elements emit from the nucleus charged particles of low mass called beta particles, which are identical to electrons. Beta radiation has medium penetrating power, between that of alpha and gamma radiation, and may be stopped by light metal such as aluminium.

Deterministic effect   The dose-dependent radioactive effect on a biological entity such as a human body. One kind of deterministic effect is radiation sickness, an often-fatal effect of exposure to a large dose of radioactivity.

D-notice   A secret government request to senior media representatives not to publish certain specified details about defence- or security-related activities. The D-notice system was adopted in Australia in 1952. D-notices were decided by

315

the Defence, Press and Broadcasting Committee administered by the Department of Defence and made up of senior government and media representatives.

Dose   The amount of energy delivered to a mass of material by ionising radiation passing through it.

Dose equivalent   Different kinds of radiation, such as gamma or alpha, have different biological effects. For example, for the same absorbed dose, alpha radiation will produce more effects than gamma radiation. The dose equivalent is measured in sieverts.

Dosimeter   A device, instrument or system used to measure or evaluate a dose of radiation. Two types of personal dosimeters were used at Maralinga by personnel entering radiation areas during the tests: quartz fibre electrometers and film badges.

Fallout   The descent to the earth's surface of particles contaminated with radioactivity, following the dispersion of radioactive material into the atmosphere by nuclear explosion. The term is applied both to the process and, in a collective sense, to the particulate matter.

Feather beds   Large metal frameworks used to hold the simulated warheads before detonation in the Vixen B safety trials held at the Taranaki firing pads at Maralinga.

Film badge   A plastic holder containing a piece of film similar to a dental x-ray film and worn by personnel at a nuclear test. Radiation exposes the film. After a nuclear test, the film is developed, and the degree of darkening apparent is a measure of the radiation dose received. The film holder usually contains metal filters to enable discrimination between different types of radiation.

Fission   The process in which the nucleus of a heavy element such as uranium or plutonium splits into two nuclei of lighter elements, accompanied by the release of substantial amounts of energy.

Forward area   The restricted zone within which the major bomb trials and minor radiological experiments took place at the British nuclear tests sites.

Fusion   The process in which the nuclei of light elements such as hydrogen (particulalry its isotopes deuterium or tritium) combine to form the nucleus of a heavier element, accompanied by the release of substantial amounts of energy.

Gamma radiation   Penetrating electromagnetic radiation emitted from the nucleus of radioactive elements. This form of radiation is most readily measured by monitoring equipment such as film badges and dosimeters.

Half-life   The time in which the activity of a radioactive species will decline to half its initial value by radioactive decay. For example, plutonium-239 has a half-life of 24 400 years, so it takes 24 400 years for half of its radioactivity to decay, then another 24 400 years for half of the remaining radiation to decay, and so on. The half-life of a radioactive species is a characteristic property of that species.

Health physics   The science of human health and radiation exposure – a branch of medical science devoted to radiation safety.

Ionising radiation   Radiation that integrates with matter to add electrons to or remove electrons from the atoms of the material absorbing it, producing electrically charged (positive or negative) atoms called ions.

Isotopes   Forms of the same element whose nuclei contain different numbers of neutrons and therefore have different mass numbers. Isotopes of an element have nearly identical chemical properties but differ in their nuclear properties. For instance, some isotopes of an element may be radioactive, and others not.

Major trials   Atomic tests conducted at Monte Bello Islands, Emu Field and Maralinga in Australia that involved detonating a complete atomic bomb, resulting in a mushroom cloud.

Minor trials   Hundreds of tests conducted at Emu Field and Maralinga in Australia that involved examining how radioactive materials and atomic weaponry would behave under various conditions such as fire or conventional explosion.

Neutron   A nuclear particle with no electric charge (neutral) and a mass approximately equal to or slightly greater than that of a proton. Neutrons are present in all atoms except those of the lightest isotope of hydrogen.

Nuclides   Species of atoms having a specified number of protons and neutrons in their nuclei. Radionuclides are the radioactive forms of nuclides. They are often expressed as, for example, ^{239}Pu, which shows in numerical form the number of neutrons combined with the number of protons (in this case 145 neutrons and 94 protons) and hence the form of isotope.

Operation Brumby   A clean-up operation, more extensive than Operation Hercules, mounted by the AWRE at the Maralinga test range between April and July of 1967. Operation Brumby was considered by the AWRE to be the final clean-up before departing the atomic weapons test site permanently.

Operation Hercules   A clean-up operation mounted by the AWRE at the Maralinga test range between August and November 1964. This clean-up was intended as a temporary measure to allow a reduction in range staff to a care and maintenance level while long-term decisions were made about its future use.

Operation RADSUR   A RADiological SURvey of both the Maralinga and the Emu Field atomic test sites carried out by the AWRE in October and November 1966. RADSUR was used as the basis for Operation Brumby.

Plumes   Clouds of radioactive material from an explosion, as well as the visible fallout after the radioactive material is carried back to the ground. The scientists on the 1984 field trip to Maralinga could detect plumes in the form of elongated hand-shapes on the ground from each detonation of the Vixen B experiments in the 1960s, because the plutonium carried back down still sat close to the surface.

Plutonium (Pu)   A dense, silvery radioactive element that does not occur naturally but is made in a reactor by irradiating uranium with neutrons. It was first produced in 1940. Plutonium has 13 known isotopes, of which plutonium-239 has the longest half-life (24 400 years). A fissile material, plutonium-239 can be used as the core of a nuclear weapon.

Quartz fibre electrometer   Dosimeters worn in the pocket like pens and read by looking through a lens to observe the position of a quartz fibre against a scale.

Radioactivity   The property of certain radionuclides of spontaneously emitting particles and/or x-ray or gamma ray radiation, or of undergoing spontaneous fission. The rate of decay is specific to a given species of radionuclide and cannot be changed by known physical or chemical processes.

Radionuclides   Radioactive nuclides.

Sievert   The unit of biological absorption of ionising radiation, expressed as dose equivalent. A millisievert is one one-thousandth of a sievert. At the time of the British nuclear tests in Australia the standard measurement of dose equivalence was the rem.

Stochastic effects   Medical conditions associated with ionising radiation, such as cancer or hereditary illness, induced at random but with no threshold radiation dose, for which probability (but not severity) increases with increasing doses of radiation. These effects may show up many years after exposure. Because stochastic effects can occur in individuals who have not been exposed to radiation above background levels, it is impossible to determine for certain whether an occurrence of cancer or genetic damage was due to a specific exposure.

Thermonuclear weapon   A nuclear device that relies on raising the temperature of a mixture of deuterium and tritium nuclei to above 10 million degrees Celsius, at which point nuclear fusion reactions occur. This type of weapon is also known as a hydrogen bomb.

Warhead   The explosive head of a bomb.

Yield   The amount of energy generated by a nuclear explosion, usually expressed in kilotonnes (for fission devices) or megatonnes (for fusion devices). A kilotonne is equivalent to 1000 tonnes of TNT, and a megatonne is equivalent to 1 million tonnes of TNT.

References

PROLOGUE

Arnold, Lorna & Smith, Mark, *Britain, Australia and the Bomb: The Nuclear Tests and Their Aftermath*, 2nd edn, Palgrave Macmillan, New York, 2006.

Beale, Howard, Minister for Supply, 'Atomic Tests in Australia', top-secret Cabinet briefing document, submission no. 73, Canberra, 11 August 1954, Malone files.

Curr, EM, 'Port Essington', in *Australian Race: Its Origin, Languages, Customs, Place of Landing in Australia, and the Routes by Which It Spread Itself over That Continent*, John Ferres, Government Printer, 1887.

Macintyre, Stuart, *A Concise History of Australia*, 2nd edn, Cambridge University Press, Port Melbourne, 2004.

Maloney, Sean M, *Learning to Love the Bomb: Canada's Nuclear Weapons during the Cold War*, Potomac Books, 2007.

Milliken, Robert, *No Conceivable Injury*, Penguin, Ringwood, 1986.

'Nuclear colonialism', Healing Ourselves and Mother Earth, <www.h-o-m-e.org/nuclear-colonialism.html>. Accessed 23 August 2014.

Research and Development Branch, Department of Supply, minutes of meeting, Swanston Street, Melbourne, 25 November 1953, National Archives of Australia (NAA): A6456, R145/011.

Royal Commission into British Nuclear Tests in Australia, *Conclusions and Recommendations* (JR McClelland, J Fitch, WJA Jones), AGPS, Canberra, 1985.

Royal Commission into British Nuclear Tests in Australia, *Report* (JR McClelland, President), 2 vols, AGPS, Canberra, 1985.

1 MARALINGA BURIED, UNCOVERED

Anderson, Ian, 'Britain's dirty deeds at Maralinga', *New Scientist*, 12 June 1993, pp. 12–13.

Arnold, Lorna & Smith, Mark, *Britain, Australia and the Bomb: The Nuclear Tests and Their Aftermath*, 2nd edn, Palgrave Macmillan, New York, 2006.

Australia, Senate, *Debates*, questions without notice (to Peter Walsh from Senator Graham Maguire), plutonium residues, Maralinga, 31 May 1984, p. 2227.

Beadell, Len, *Blast the Bush*, Rigby Limited, Adelaide, 1967.

Bernstein, Jeremy, *Plutonium: A History of the World's Most Dangerous Element*, UNSW Press, Sydney, 2001.

Blakeway, Denys & Lloyd-Roberts, Sue, *Fields of Thunder: Testing Britain's Bomb*, George Allen & Unwin, London, 1985.

Brown, Paul, 'British nuclear testing in Australia: performing the Maralinga experiment through verbatim theatre', *Journal and Proceedings of the Royal Society of New South Wales*, vol. 139, 2006.

Burns, Peter & Williams, Geoff, tape recording of interview with Ian Anderson, 1993 (exact date unknown), Anderson files.

Burns, Peter & Williams, Geoff, interview with author, ARPANSA, Melbourne, 15 April 2004.

Cawte, Alice, *Atomic Australia 1944–1990*, UNSW Press, Kensington, 1992.

Connor, Steve, 'WA atom blast was far bigger, UK says', *Sydney Morning Herald (SMH)*, 26 May 1984, p. 3.

Department of Education, Science and Training, 'Rehabilitation of Former Nuclear Tests Sites at Emu and Maralinga (Australia) 2003' (MARTAC Report), Commonwealth of Australia, Canberra, 2002.

Lokan, KH, Head of ARL, evidence to Joint Committee on Public Works, 23 February 1995, in minutes of evidence relating to Maralinga Rehabilitation Project, Parliament of the Commonwealth of Australia, 1995.

MARTAC Report, see Department of Education, Science and Training.

Moroney, John, annotated draft of Ian Anderson's *New Scientist* story, unpublished, 2 June 1993, Anderson files.

Ophel, TR, 'Sir Ernest William Titterton', Obituary, *ANU Reporter*, 23 February 1990, p. 4.

Pearce, Noah, 'Final Report on Residual Radioactive Contamination of the Maralinga Range and the Emu Site', AWRE report no. 01–16/68 (Pearce Report), UK Atomic Energy Authority, January 1968. Edited version released by Australian Department of National Development, May 1979. Full version tabled by Australian Senate, May 1984.

Pearce Report, see Pearce, Noah.

Resture, Jane, 'Return to Maralinga', Jane Resture's Oceania Page, 2012, <www.janesoceania.com/christmas_about1/index.htm>. Accessed 23 August 2014.

Royal Commission into British Nuclear Tests in Australia, *Conclusions and Recommendations* (JR McClelland, J Fitch, WJA Jones), AGPS, Canberra, 1985.

Royal Commission into British Nuclear Tests in Australia, *Report* (JR McClelland, President), 2 vols, AGPS, Canberra, 1985.

Stanton, John, 'Plutonium dumps a risk "for thousands of years"', *The Australian*, 25 May 1984, p. 2.

Toohey, Brian, 'Killen warns on plutonium pile', *Australian Financial Review*, 5 October 1978.

Toohey, Brian, 'Plutonium on the wind: the terrible legacy of Maralinga', *National Times*, 4–10 May 1984, pp. 3–5.

United Nations Office for Disarmament Affairs, 'Treaty on the Non-Proliferation of Nuclear Weapons: text of the treaty', <disarmament.un.org/treaties/t/npt/text>. Accessed 28 February 2016.

Walsh, Peter, statement, Senate, 4 May 1984, reproduced in *Australian Foreign Affairs Record*, vol. 55, no. 5, May 1984, p. 486.

Walsh, Peter, media release, 15 May 1984, reproduced in *Australian Foreign Affairs Record*, vol. 55, no. 5, May 1984, p. 547.

Walsh, Peter, statement, Senate, 7 June 1984, reproduced in *Australian Foreign Affairs Record*, vol. 55, no. 6, June 1984, p. 636.

Walsh, Peter, letter to Howard Conkey, *Canberra Times*, 28 June 1984.

Walsh, Peter, *Confessions of a Failed Finance Minister*, Random House Australia, Milsons Point, 1995.

Wick, OJ, *Plutonium Handbook: A Guide to the Technology*, American Nuclear Society, Grange Park, 1980.

2 BRITAIN'S STEALTHY MARCH TOWARDS THE BOMB

'Alan Nunn May, 91, pioneer in atomic spying for Soviets', *New York Times*, 25 January 2003.

Allen, Christian, 'Atom spy Klaus Fuchs jailed', *History Today*, LookSmart, 2000, <www.historytoday.com/christian-allen/atom-spy-klaus-fuchs-jailed>. Accessed 19 November 2015.

Aylen, Jonathan, 'First waltz: development and deployment of Blue Danube, Britain's post-war atomic bomb', *International Journal for the History of Engineering and Technology*, vol. 85, no. 1, January 2015, pp. 31–59.

Cathcart, Brian, 'Obituary: Theodore Hall', *Independent*, 12 November 1999, <www.independent.co.uk/arts-entertainment/obituary-theodore-hall-1125267.html>. Accessed 9 December 2010.

Fitzgerald, EM, 'Allison, Attlee and the bomb: views on the 1947 British decision to build an atom bomb', *RUSI Journal*, vol. 122, issue 1, 1977.

Foulkes, JN & Thompson, DS, 'A Biological Survey of the Maralinga Tjuratja Lands, South Australia, 2001–2007', Science Resource Centre, Information, Science and Technology Directorate, Department for Environment and Heritage, South Australia, December 2008.

Goodman, Michael S, 'The grandfather of the hydrogen bomb? Anglo-American intelligence and Klaus Fuchs', *Historical Studies in the Physical and Biological Sciences*, vol. 34, no. 1, 2003, pp. 1–22.

Gowing, Margaret, *Britain and Atomic Energy 1939–1945*, Macmillan, London, 1964.

Gowing, Margaret, 'The men', in Margaret Gowing with Lorna Arnold, *Independence and Deterrence: Britain and Atomic Energy, 1945–1952*, Palgrave Macmillan, 1974, Chapter 13.

Gowing, Margaret, 'James Chadwick and the atomic bomb', *Notes and Records of the Royal Society of London*, vol. 47, no. 1, January 1993.

Grabosky, PN, 'A toxic legacy: British nuclear weapons tests', in *Wayward Governance: Illegality and Its Control in the Public Sector*, Australian Institute of Criminology, Canberra, 1989, Chapter 16.

Hennessy, Peter, 'Cabinets and the bomb', Inaugural Michael Quinlan Lecture, UK House of Lords, 2 February 2011.

Meade, Roger & Meade, Linda (eds), 'Klaus Fuchs: the second confession', Los Alamos National Laboratory Historical Document, 2014, <permalink.lanl.gov/object/tr?what=info:lanl-repo/lareport/LA-UR-14-27960>. Accessed 20 July 2015.

Newton, JO, 'Ernest William Titterton 1916–1990', *Historical Records of Australian Science*, vol. 9, no. 2, 1992.

Oppenheimer, J Robert, 'Now I am become death', video recording, Atomic Archive, <www.atomicarchive.com/Movies/Movie8.shtml>. Accessed June 2015.

Paterson, Robert H, *Britain's Strategic Nuclear Deterrent: From before the V-Bomber to beyond Trident*, Routledge, 2012.

'Plutonium and Aldermaston – an historical account', Federation of American Scientists, 2000, <fas.org/news/uk/000414-uk2.htm>. Accessed 19 November 2015.

Schrafstetter, Susanna, "'Loquacious … and pointless as ever"? Britain, the United States and the United Nations negotiations on international control of nuclear energy, 1945–48', *Contemporary British History*, vol. 16, no. 4, 2002, pp. 87–108.

Spinardi, Graham, 'Aldermaston and British nuclear weapons development: testing the "Zuckerman thesis"', *Social Studies of Science*, vol. 27, no. 4, 1997, pp. 547–582.

'Spy's deathbed confession: atom physicist tells how secrets given to Soviet Union',
    Guardian, 27 January 2003.
Symonds, JL, *A History of British Atomic Tests in Australia*, AGPS, Canberra, 1985.
Truman, Harry S, 'Statement announcing the atomic bombing of Hiroshima, August 6,
    1945', American Experience, <www.pbs.org/wgbh/americanexperience/features/
    primary-resources/truman-hiroshima>. Accessed 11 October 2015.
WikiLeaks, 'United Kingdom atomic weapons program: the full Penney report (1947)',
    <wikileaks.org/wiki/United_Kingdom_atomic_weapons_program:_The_full_
    Penney_Report_%281947%29>. Accessed 20 July 2015.

3 MONTE BELLO AND EMU FIELD

Acaster, Ray, 'Worlds apart: atom bombs and traditional land use in South Australia',
    Limina, vol. 1, 1995.
Arnold, Lorna & Smith, Mark, *Britain, Australia and the Bomb: The Nuclear Tests and Their
    Aftermath*, 2nd edn, Palgrave Macmillan, New York, 2006.
'Atom bombs in our arid lands', *Sunday Herald*, 4 October 1953, p. 2.
'Atomic bomb exploded in Monte Bello Islands', *West Australian*, 4 October 1952,
    reproduced in FK Crowley, *Modern Australian Documents 1939–1970*, vol. 2, Wren
    Publishing, Melbourne, 1973.
'Atomic bombs, uranium and Australia's future', *SMH*, 8 October 1953, p. 2.
Atomic Weapons Research Establishment Report No. T1/54 – Operation Hurricane
    Director's Report – Scientific Data Obtained at Operation Hurricane –
    November 1954, Copy 36, 1954, NAA: A6454, Z5A.
'The atom's challenge to humanity', *SMH*, 16 October 1953, p. 2.
Attendance of Australian Defence Scientific Adviser at Monte Bello, extract from draft
    reply to letter of 27 August 1952 from UK High Commissioner's Office, Canberra,
    1952, NAA: A6456, R091/003.
'Australian Armed Services Have Played Important Role in Monte Bello Atomic Test
    Plans: Secret Preparations Have Been Going on for More than 24 Months', 1952,
    NAA: A816, 3/301/539A.
Beadell, Len, *Blast the Bush*, Rigby Limited, Adelaide, 1967.
Beadell, Len, 'Too long in the bush', transcript of talk to Rotary Convention,
    Shepparton, Victoria, 1991, private recording.
Beale, Howard, 'Future Atomic Tests in Australia', letter to Robert Menzies,
    12 November 1953, NAA: A6456, R075/062.
Beale, Howard, Minister for Supply, 'Atomic Tests in Australia', top-secret Cabinet
    briefing document, submission no. 73, Canberra, 11 August 1954, Malone files.
Beale, Howard, *This Inch of Time: Memoirs of Politics and Diplomacy*, Melbourne University
    Press, Melbourne, 1977.
Blakeway, Denys & Lloyd-Roberts, Sue, *Fields of Thunder: Testing Britain's Bomb*, George
    Allen & Unwin, London, 1985.
Brown, AS, Secretary, Prime Minister's Department, 'Hurricane', letter to Ben Cockram,
    Office of the High Commission for the UK, Canberra, 19 September 1952, NAA:
    A6456, R091/003.
Clearwater, John & O'Brien, David, '"Oh lucky Canada": radioactive polar bears; the
    proposed testing of British nuclear weapons in Canada', *Bulletin of the Atomic
    Scientists*, July–August 2003.

References

Cockram, Ben, Office of the High Commissioner for the UK, Canberra, letter to AS Brown, Prime Minister's Department, Canberra, 29 September 1952, NAA: A6456, R091/003.

Collins, Vice Admiral Sir John A, First Naval Member, Chief of Naval Staff, Navy Office, Melbourne, letter marked 'Secret and personal' to Sir Frederick Shedden, Secretary, Department of Defence, 22 August 1952, NAA: A6456, R091/003.

Department of Defence, 'Atomic Test – Background Articles', minute paper, 26 August 1952, NAA: A816, 3/301/539A.

Department of Supply, notes of interdepartmental committee meeting to consider decision no. 69 (PM) Cabinet minute dated 26 August 1954, 6 September 1954, Malone files.

Durie, R, Acting Secretary, Cabinet, Cabinet minute, decision no. 69, Prime Minister's Committee, Canberra, 26 August 1954, Malone files.

'For Press: Atomic Weapon Test', media release, 13 August 1952, NAA: A816, 10/301/129.

Gowing, Margaret with Arnold, Lorna, *Independence and Deterrence: Britain and Atomic Energy, 1945–1952*, Palgrave Macmillan, 1974.

Grabosky, PN, 'A toxic legacy: British nuclear weapons tests', in *Wayward Governance: Illegality and Its Control in the Public Sector*, Australian Institute of Criminology, Canberra, 1989, Chapter 16.

A 'Herald' Special Reporter who watched the atomic explosion on the Woomera Range from 15 miles away, 'Atom explosion success: dawn blast at Woomera', *SMH*, 16 October 1953, p. 1.

'His responsibility!', *Daily Mirror* (Sydney), 2 October 1953, p. 2.

Ison, Erin, 'Bombs, "reds under the bed" and the media: the Menzies government's manipulation of public opinion, 1949–1957', *Humanity*, University of Newcastle, 2008.

Maralinga press visit, June 1956, NAA: A6456, R087/135.

Martin, AW, *Robert Menzies: A Life*, vol. 2, *1944–1978*, Melbourne University Press, Melbourne, 1999.

McClellan, Peter, 'Who is telling the truth? Psychology, common sense and the law', Local Courts of New South Wales Annual Conference, August 2006.

McKnight, AD, Prime Minister's Department, letter to George Davey, Office of the High Commissioner for the UK, Canberra, 30 September 1952, NAA: A6456, R096/006.

Milliken, Robert, *No Conceivable Injury*, Penguin, Ringwood, 1986.

'Months of hard, lonely work paved way for atomic explosion', *SMH*, 16 October 1953, p. 4.

Morgan, Sue, 'Britain's star witness sheds little light', *SMH*, 2 February 1985, p. 12.

Moroney, John, 'Aide Memoire on Health Control and Surveillance in the Nuclear Tests in Australia', unpublished briefing document, ARL, 24 December 1992.

Nason, David, 'Owners to reclaim Maralinga bomb site', *The Australian*, 10 November 2009, p. 7.

Norris, Robert Standish, *Questions on the British H-Bomb*, Natural Resources Defense Council, 22 June 1992, p. 3.

Obituary for Lord Penney, *The Times*, 6 March 1991.

'Other "minor" tests', *SMH*, 28 October 1953, p. 1.

Penney, William, statement, Royal Commission, 1984–1985, NAA: A6449, 2 (M–Z).

Pincher, Chapman, 'Could be dropped by R.A.F. plane', *SMH*, 16 October 1953, p. 1.

Porter, David, 'Atomic bomb lessons in the desert', *SMH*, 22 June 1988, p. 2.

'Put method in atomic check', *Herald* (Melbourne), 22 June 1956, clippings package, NAA: A5954, 2167/6.

'The R.A.N. Prepared a New Chart for the Monte Bello Atomic Test', 1952, NAA: A816, 3/301/539A.

'Roads and Landings Had to Be Built for Monte Bello Atomic Weapon Test: Heavy Equipment Taken There by Land and Sea', 1952, NAA: A816, 3/301/539A.

Royal Commission into British Nuclear Tests in Australia, *Conclusions and Recommendations* (JR McClelland, J Fitch, WJA Jones), AGPS, Canberra, 1985.

Royal Commission into British Nuclear Tests in Australia, *Report* (JR McClelland, President), 2 vols, AGPS, Canberra, 1985.

'Second A-blast successful: trial series ends', *SMH*, 28 October 1953, p. 1.

Secretary of State for Commonwealth Relations, top-secret telegram to High Commissioner for the UK, Australia, 2 April 1955, National Archives of the UK, FCO/1/8.

Smith, Joan, *Clouds of Deceit: The Deadly Legacy of Britain's Bomb Tests*, Faber & Faber, London, 1985.

Solomon, Colonel GD, Range Commander, report on the Maralinga range, 9 December 1959, NAA: A6456, R047/022.

Special Correspondent in London, 'How an atomic bomb explodes', *SMH*, 16 October 1953, p. 2.

Staff Correspondent, 'Long-range risks in atomic tests?', *SMH*, 10 October 1953, p. 2.

Staff Correspondent, 'Public safety and the Woomera tests', *SMH*, 9 October 1953, p. 2.

'Substance of Communication Dated 18th October 1952 from High Commissioner for the United Kingdom, Canberra', 1952, NAA: A6456, R021/001 PART 36.

Symonds, JL, 'British Atomic Tests in Australia Chronology of Events: 1950–1968', 1984–1985, NAA: A6456/3, R023/003.

Symonds, JL, *A History of British Atomic Tests in Australia*, AGPS, Canberra, 1985.

'Threat to 3 towns', *Sun* (Sydney), 21 June 1956, p. 1.

Titterton, EW, 'The race for atomic arms', *Sunday Herald*, 4 October 1953, p. 2.

4 MUSHROOM CLOUDS AT MARALINGA

Arnold, Lorna & Smith, Mark, *Britain, Australia and the Bomb: The Nuclear Tests and Their Aftermath*, 2nd edn, Palgrave Macmillan, New York, 2006.

AWTSC, minutes of 21st meeting, Department of Supply, Swanston St, Melbourne, 19 July 1957, NAA: A6456, R120/137 PART 1.

Beadell, Len, 'Too long in the bush', transcript of talk to Rotary Convention, Shepparton, Victoria, 1991, private recording.

Beale, Howard, Minister for Supply, memorandum, Canberra, 4 May 1955, National Archives of the UK, FCO 1 358/2099.

Beale, Howard, 'Why We Hold A-Tests in Australia', 6 August 1956, NAA: A6456, R047/011.

Beale, Howard, 'Why Australia provides the site for atomic tests', *SMH*, 13 August 1956, p. 2.

Beale, Howard, transcript of joint press conference by Hon. Howard Beale QC MP, Minister for Supply & Sir William Penney, KBE FRS DSc, Director, AWRE, 14 August 1956, NAA: A6456, R087/135.

Beale, Howard, media statement, 18 September 1956, NAA: A6456, R047/011.

Beale, Howard, 'Press Reports on A-Test: Statement by the Minister for Supply, the

References

Hon. Howard Beale Q.C.', media release, Canberra, 26 September 1956, NAA: A6456, R124/025.

Beale, Howard, *This Inch of Time: Memoirs of Politics and Diplomacy*, Melbourne University Press, Melbourne, 1977.

Buckley, Dan, British Army major (retd), statement, Royal Commission, 1984–1985, NAA: A6449, 1 (A–L).

Cook, Melbourne, cable on press arrangements to Herington, Adelaide, 11 September 1956, NAA: A6456, R047/011.

Department of Supply, 'Maralinga Project – Review of Australian Support', briefing document for media, 1956, NAA: A6456, R087/205.

Department of Supply, 'New A-Tests at Maralinga Next Month', media release, 29 August 1957, NAA: A6456, R087/090.

Fairley, Douglas, 'D notices, official secrets and the law', *Oxford Journal of Legal Studies*, vol. 10, no. 3, Autumn 1990, p. 432.

'Freedom safeguard: more atomic tests', *Sun* (Sydney), 11 October 1956, p. 40.

Keane, John, 'Maralinga's afterlife', *The Age*, 11 May 2003.

Kendall, GM, Muirhead, CR, Darby, SC, Doll, R, Arnold, L & O'Hagan, JA, 'Epidemiological studies of UK test veterans: I. General description', *Journal of Radiological Protection*, vol. 24, 27 August 2004, p. 203.

'Labor opposes atomic tests', *Advertiser*, 21 June 1956, clippings package, NAA: A6456, R058/006.

'Labor will stop weapons test', *Canberra Times*, 25 June 1956, clippings package, NAA: A6456, R058/006.

'Latest on the bomb!', *Sun* (Sydney), 25 September 1956, p. 1.

'List of Passengers Who Will Travel to Maralinga in DC3 Guineas Aircraft on Tuesday, 19 June, Departing West Beach at 7.15 am', 1956, NAA: A6456, R087/135.

Lowe, Peter, British Army colonel (retd), statement, Royal Commission, 1984–1985, NAA: A6449, 1 (A–L).

Lowe, Peter, British Army colonel (retd), Royal Commission, 3–30 January 1985, NAA: A6448, 9.

Maralinga press visit, June 1956, NAA: A6456, R087/135.

Maralinga Rehabilitation Project, Parliament of the Commonwealth of Australia, minutes of evidence, 1995.

McClelland, James, transcript of interview with Robin Hughes, Australian Biography Project, 25 January 1995, <www.australianbiography.gov.au/subjects/mcclelland/interview6.html>. Accessed 20 November 2015.

Milliken, Robert, *No Conceivable Injury*, Penguin, Ringwood, 1986.

Ministry of Defence (UK), draft telegram prepared 29 April 1955, received by Foreign and Commonwealth Office 4 May 1955, National Archives of the UK, FCO 1/8.

'Misgivings over atomic tests', *The Age*, 22 June 1956, clippings package, NAA: A5954, 2167/6.

Moroney, John, 'Aide Memoire on Health Control and Surveillance in the Nuclear Tests in Australia', unpublished briefing document, ARL, 24 December 1992.

Moroney, John, 'Aide Memoire on the Nuclear Tests in Australia in the Context of the British Weapons Development Program', unpublished briefing document, ARL, 24 December 1992.

'The mushroom cloud', Atomic Archive, <www.atomicarchive.com/Effects/effects9.shtml>. Accessed 27 September 2015.

Newton, JO, 'Ernest William Titterton 1916–1990', *Historical Records of Australian Science*, vol. 9, no. 2, 1992, p. 178.

O'Connor, FA, letter to Iyer Jehu, 9 November 1956, NAA: A6456, R030/075.

O'Connor, FA, 'Expenses in Connection with the Visit of the Press Party to Witness an Atomic Explosion at Maralinga, 1956', minute paper, [1957], NAA: A6456, R029/249.

O'Connor, FA, Secretary, Department of Supply, memorandum to Australian Newspaper Proprietors' Association, 3 September 1957, NAA: A6456, R087/090.

O'Connor, FA, teleprinter message to Minister for Supply Howard Beale, quoting message from William Penney, 30 August 1956, NAA: A6456, R029/249.

Parkinson, Alan, *Maralinga: Australia's Nuclear Waste Cover-Up*, ABC Books, Sydney, 2007.

Penney, William, *Guest of Honour* program, ABC, 2 September 1956, transcript of broadcast, NAA: A6456, R030/074.

Powell, RR, UK Ministry of Defence, letter to JM Wilson, UK Ministry of Supply, 9 February 1955, National Archives of the UK, FCO 1/8.

'Press Reaction to Atomic Tests', [no author, likely to be a Department of Supply bureaucrat, undated], NAA: A6456, R047/011.

'Put method in atomic check', *Herald* (Melbourne), 22 June 1956, clippings package, NAA: A5954, 2167/6.

Rees, Margaret, 'Documents confirm soldiers were exposed to nuclear tests in Australia', World Socialist, 9 July 2001, <www.wsws.org/articles/2001/jul2001/mara-j09.shtml>. Accessed 18 December 2009.

'Reward for a good boy', *Sun-Herald*, 19 January 1958, clipping in biographical file for Howard Beale, National Library of Australia.

Royal Commission into British Nuclear Tests in Australia, *Report* (JR McClelland, president), 2 vols, AGPS, Canberra, 1985.

Solomon, Colonel GD, Range Commander, report on the Maralinga range, 9 December 1959, NAA: A6456, R047/022.

Special Correspondent, 'Australian atomic test site', *SMH*, 17 May 1955, p. 2.

Staff Correspondent, 'Safety of A-weapon tests at Maralinga', *SMH*, 28 June 1956, p. 2.

Tait, Gordon, Chief of Bureau, Associated Press, letter to Reg Harris, Department of Supply, 15 October 1957, NAA: A6456, R087/090.

Titterton, EW, *Facing the Atomic Future*, Macmillan, London, 1956.

Titterton, EW, 'Answers to questions on atomic tests', *The Age*, 15 May 1956, p. 2.

Titterton, EW, 'Some questions and answers on latest atom tests', *SMH*, 15 May 1956, p. 2.

Titterton, EW, 'Why Australia is atom testing ground', *The Age*, 16 May 1956, p. 2.

Titterton, EW, 'Why Australia is preferred as an atom testing ground', *SMH*, 16 May 1956, p. 2.

Titterton, EW, 'After atomic bomb tests … how dangerous is the mushroom cloud?', *The Age*, 19 July 1956, p. 2.

UK Atomic Weapons Trial Executive, minutes of meeting, 10 October 1956, quoted in Royal Commission into British Nuclear Tests in Australia, *Report* (JR McClelland, President), vol. 2, AGPS, Canberra, 1985, pp. 494–495.

5 VIXEN B AND OTHER 'MINOR TRIALS'

Admiralty, War Office, Air Ministry and Press Committee, Department of Defence confidential briefing document, [1952], NAA: A5954, 1956/6.

References

Arnold, Lorna, *A Very Special Relationship: British Atomic Weapon Trials in Australia*, Her Majesty's Stationery Office, London, 1987.

Arnold, Lorna & Smith, Mark, *Britain, Australia and the Bomb: The Nuclear Tests and Their Aftermath*, 2nd edn, Palgrave Macmillan, New York, 2006.

Australian Nuclear Veterans Association, South Australia & Maralinga and Monte Bello Islands Ex-Servicemen's Association, submission, quoted in 'Protection was "inadequate"', *SMH*, 19 September 1995.

Australian Participants in British Nuclear Tests (Treatment) Bill 2006, bills digest no. 31, 2006–07, Parliament of Australia Parliamentary Library.

Barnaby, Frank, 'The management of radioactive wastes and the disposal of plutonium', paper at Medical Association for Prevention of War Conference, 2000.

Blakeway, Denys & Lloyd-Roberts, Sue, *Fields of Thunder: Testing Britain's Bomb*, George Allen & Unwin, London, 1985.

'Broken arrows: nuclear weapons accidents', Atomic Archive, <www.atomicarchive. com/Almanac/Brokenarrows_static.shtml>. Accessed 13 March 2016.

Brooking, PWB, letter to Sir William Cook, 29 August 1958, NAA: A6455, RC386.

Brooking, PWB, letters to Director, AWRE, 23 & 29 September 1958, NAA: A6455, RC386.

Brown, Paul, 'British nuclear testing in Australia: performing the Maralinga experiment through verbatim theatre', *Journal and Proceedings of the Royal Society of New South Wales*, vol. 139, 2006.

Bryant, AR, 'Maralinga Minor Trials in Relation to a Ban on Nuclear Testing', 29 August 1958, NAA: A6455, RC386.

Burns, Peter & Williams, Geoff, interview with author, ARPANSA, Melbourne, 15 April 2004.

Carter, Raymond Frank, statement, Royal Commission, 1984–1985, NAA: A6449, 1 (A–L).

Costar, NE, Office of the High Commissioner for the UK, Canberra, letter to MC Timbs, Prime Minister's Department, 20 October 1960, NAA: A6456, R150/001.

Cross, Roger, *Fallout: Hedley Marston and the British Bomb Tests in Australia*, Wakefield Press, Kent Town, 2001.

Cross, Roger & Hudson, Avon, *Beyond Belief: The British Bomb Tests; Australia's Veterans Speak Out*, Wakefield Press, Kent Town, 2005.

Crouch, PC, Robotham, FJP & Williams, GA, 'Submission to the Senate Foreign Affairs, Defence and Trade Committee on the Australian Participants in British Nuclear Tests (Treatment) Bill 2006; and the Australian Participants in British Nuclear Tests (Treatment) (Consequently Amendments and Transitional Provisions) Bill 2006', 25 October 2006.

Crowley, FK, *Modern Australian Documents 1939–1970*, vol. 2, Wren Publishing, Melbourne, 1973.

Department of Education, Science and Training, 'Rehabilitation of Former Nuclear Tests sites at Emu and Maralinga (Australia) 2003' (MARTAC Report), Commonwealth of Australia, Canberra, 2002.

Draft press statements 1 & 2 on activities at the Maralinga range, 1960, NAA: A6456/3, R124/025.

Fairley, Douglas, 'D notices, official secrets and the law', *Oxford Journal of Legal Studies*, vol. 10, no. 3, Autumn 1990.

Grabosky, PN, 'A toxic legacy: British nuclear weapons tests', in *Wayward Governance: Illegality and Its Control in the Public Sector*, Australian Institute of Criminology, Canberra, 1989, Chapter 16.

Gregory, Shaun, *The Hidden Cost of Deterrence: Nuclear Weapons Accidents*, Brassey's (UK), London, 1990.

Griffiths, LF, Senior Works Supervisor, Department of Works, 'Taranaki – Vixen "B" – Site Meeting', letter to Director of Works, Adelaide, 1 August 1960, NAA: A6456, R129/002.

Knott, JL, Acting Secretary, Department of Supply, letter to Allen Fairhall, 29 July 1959, NAA: A6456, R105/001.

Lewis, Richard, Morris, Annette & Oliphant, Ken, 'Tort personal injury claims statistics: is there a compensation culture in the United Kingdom?', Cardiff Law School, Cardiff University, <www.law.cardiff.ac.uk/research/pubs/repository/1445.pdf>. Accessed 23 March 2016.

Lokan, KH & Williams, GA, 'Maralinga – Radiological Safety for APS Personnel', first prepared June 1988, revised January 1995, attachment to PC Crouch, FJP Robotham & GA Williams, 'Submission to the Senate Foreign Affairs, Defence and Trade Committee on the Australian Participants in British Nuclear Tests (Treatment) Bill 2006; and the Australian Participants in British Nuclear Tests (Treatment) (Consequently Amendments and Transitional Provisions) Bill 2006', 25 October 2006.

Macintyre, Stuart, *A Concise History of Australia*, 2nd edn, Cambridge University Press, Port Melbourne, 2004.

Maher, Laurence W, 'National security and mass media self-censorship: the origins, disclosure, decline and revival of the Australian D-notice system', *Australian Journal of Legal History*, vol. 3, 1997.

Maher, Laurence W, 'H.V. Evatt and the Petrov defection: a lawyer's interpretation', *Australian Society of Labour History*, 2007, <www.historycooperative.org/proceedings/asslh2/maher.html>. Accessed 31 December 2009.

MARTAC Report, see Department of Education, Science and Training.

Martin, AW, *Robert Menzies: A Life*, vol. 2, *1944–1978*, Melbourne University Press, Melbourne, 1999.

McKnight, David, *Australian Spies and Their Secrets*, Allen & Unwin, Sydney, 1994.

McLean, Major J, NS group report on experimental measures carried out during the 1960 Vixen B trials, 1961, NAA: A6454, ZB29.

Menzies, Robert, *The Measure of the Years*, Cassell Australia, Melbourne, 1970.

Milliken, Robert, *No Conceivable Injury*, Penguin, Ringwood, 1986.

Moore, Daniel, '30 years on: Britain agrees to pay its Maralinga victims', *SMH*, 4 April 1988, p. 1.

Moroney, John, letter to Roy Pilgrim, 8 November 1963, NAA: A6456, R069/032.

Moroney, John, 'Radiological Hazards of Plutonium', memorandum, 25 June 1964, NAA: A6456, R069/032.

Moroney, John, 'Aide Memoire on the Nuclear Tests in Australia in the Context of the British Weapons Development Program', unpublished briefing document, ARL, 24 December 1992.

Obituary for Maurice Timbs, *The Australian*, 22 December 1994.

Oldbury, AE, Decontamination group report for Vixen B series 1960, 1961, NAA: A6454, ZB29.

References

Parkinson, Alan, 'The Maralinga rehabilitation project: final report', *Journal of Medicine, Conflict and Survival*, vol. 20, 2004.

Parkinson, Alan, *Maralinga: Australia's Nuclear Waste Cover-Up*, ABC Books, Sydney, 2007.

Pearce, Noah, Superintendent, Radiation Measurements and Instrumentation, AWRE, letter to John Moroney, Secretary, AWTSC, 24 July 1964, NAA: A6456, R069/032.

Pilgrim, Roy, Maralinga Experimental Programme, safety statements, 1960, 1961, 1963, NAA: A6455, RC371.

Pilgrim, Roy, 'Vixen B – Firing Safety Circuit', memorandum to AWRE Deputy Director EF Newley, 23 August 1960, National Archives of the UK, ES 1/962.

Pilgrim, Roy, letter to Ernest Titterton, 11 October 1962, National Archives of the UK, DEFE 16/854.

'Report on the Circumstances Attending the Loss of Seven Captive Balloons on the Night of 23rd/24th September 1960', 1960, NAA: A6456, R088/026.

Royal Commission into British Nuclear Tests in Australia, *Report* (JR McClelland, President), 2 vols, AGPS, Canberra, 1985.

Seldon, Robert, Lawrence Livermore Laboratory, quoted in Frank Barnaby, Plutonium and Radioactive Waste Management Policy Consultant, 'The management of radioactive wastes and the disposal of plutonium', paper at Medical Association for Prevention of War Conference, 2000.

Sherratt, Tim, 'A political inconvenience: Australian scientists at the British atomic weapons tests, 1952–53', *Historical Records of Australian Science*, vol. 6, no. 2, December 1985.

Spinardi, Graham, 'Aldermaston and British nuclear weapons development: testing the "Zuckerman thesis"', *Social Studies of Science*, vol. 27, no. 4, 1997, pp. 547–582.

Stewart, K, 'Vixen "B" 1960: Nuclear Safety Group Experimental Plan', 8 July 1960, National Archives of the UK, ES 1/962.

Symonds, JL, 'British Atomic Tests in Australia Chronology of Events: 1950–1968; The Minor Trials – Assessment and Safety Tests', 1984–1985, NAA: A6456/3, R023/003.

Symonds, JL, *A History of British Atomic Tests in Australia*, AGPS, Canberra, 1985.

Timbs, MC, letter to Prime Minister Robert Menzies, 27 September 1960, NAA: A6456/3, R124/025.

Titterton, EW, *Facing the Atomic Future*, Macmillan, London, 1956.

Toohey, Brian, 'Killen warns on plutonium pile', *Australian Financial Review*, 5 October 1978.

Walker, John R, *British Nuclear Weapons and the Test Ban 1954–1973*, Ashgate, Surrey, 2010.

Williams, Geoff, personal correspondence with author, 23 February 2016.

Woollett, SM, 'Rehabilitating Maralinga', *Clean Air and Environmental Quality*, vol. 37, issue 4, November 2003.

6 THE AUSTRALIAN SAFETY COMMITTEE

Adams, CA, 'Co 60 Pellets Found at Maralinga', top-secret memorandum to Admiral PW Brooking, 12 August 1958, National Archives of the UK, DEFE 16/144.

Arnold, Lorna & Smith, Mark, *Britain, Australia and the Bomb: The Nuclear Tests and Their Aftermath*, 2nd edn, Palgrave Macmillan, New York, 2006.

AWTSC, minutes of 3rd meeting, University of Melbourne, 28 November 1955, NAA: A6456, R120/137 PART 1.

AWTSC, minutes of 4th meeting, University of Melbourne, 10 January 1956, NAA:A6456, R120/137 PART 1.

AWTSC, minutes of 19th meeting, Commonwealth X-Ray and Radium Laboratory, Surry Place, Melbourne, 11 June 1957, NAA:A6456, R120/137 PART 1.

AWTSC, draft of journal article, 'Radioactive fallout Australia from Operation Buffalo', NAA:A6456, R055/011, later published in *Australian Journal of Science*, vol. 21, no. 3, October 1958.

Beale, Howard, 'Press Reports on A-Test: Statement by the Minister for Supply, the Hon. Howard Beale Q.C.', media release, Canberra, 26 September 1956, NAA:A6456, R124/025.

Burns, Peter & Williams, Geoff, interview with author, ARPANSA, Melbourne, 15 April 2004.

Cabinet minute, decision no. 522, 16 April 1973, NAA:A5915, 258.

Cross, Roger, *Fallout: Hedley Marston and the British Bomb Tests in Australia*, Wakefield Press, Kent Town, 2001.

Expert Committee on the Review of Data on Atmospheric Fallout Arising from British Nuclear Tests in Australia, *Report* (CB Kerr, Chair) (Kerr Report), 31 May 1984, Malone files.

Kerr Report, see Expert Committee on the Review of Data on Atmospheric Fallout Arising from British Nuclear Tests in Australia.

Knott, JL, Secretary, Department of Supply, letter to EW Titterton, 26 August 1960, NAA: 6456, R150/001.

Leonard, Zeb, 'Tampering with history: varied understandings of Operation Mosaic', *Journal of Australian Studies*, vol. 38, no. 2, May 2014, p. 219.

Martin, Professor Leslie, 'Re Monte Bello Trials – Safety of Shipping', teleprinter message to Minister for Supply, 1 June 1956, NAA:A6456, R094/009.

Martin, Professor Leslie, Chair, AWTSC, letter to Frank O'Connor, Department of Supply, 20 August 1956, NAA:A6455, RC596.

McClelland, James, *Stirring the Possum: A Political Autobiography*, Penguin, Ringwood, 1988.

McClelland, James, transcript of interview with Robin Hughes, Australian Biography Project, 25 January 1995, <www.australianbiography.gov.au/subjects/mcclelland/interview6.html>. Accessed 20 November 2015.

Menzies, Robert, letter to Athol Townley, 16 May 1955, quoted in JL Symonds, 'British Atomic Tests in Australia Chronology of Events: 1950–1968', 1984–1985, NAA:A6456/3, R023/003.

Menzies, Robert, reply to question from Mr Bruce, extract from Parliamentary Debates, 22nd Parliament, 6–8 March 1956, NAA:A6456, R094/009.

Milliken, Robert, *No Conceivable Injury*, Penguin, Ringwood, 1986.

Moroney, John, Royal Commission, 5–24 June 1985, NAA:A6448, 15.

Moroney, John, 'Aide Memoire on Health Control and Surveillance in the Nuclear Tests in Australia', unpublished briefing document, ARL, 24 December 1992.

Newton, JO, 'Ernest William Titterton 1916–1990', *Historical Records of Australian Science*, vol. 9, no. 2, 1992.

O'Connor, Frank, Department of Supply, 'Scientific Press Releases – Maralinga', cable to Minister for Supply Howard Beale, 22 August 1956, NAA:A6455, RC596.

O'Connor, Frank, Department of Supply, cable to F Hinshelwood, Department of Supply, 28 August 1956, NAA:A6455, RC596.

O'Connor, Frank, Department of Supply, memorandum to Professor Leslie Martin, AWTSC, 28 August 1956, NAA: A6455, RC596.

Ophel, Trevor & Jenkin, John, *Fire in the Belly: The First 50 Years of the Pioneer School at the ANU*, ANU, Canberra, 1996.

Penney, William, statement, Royal Commission, 1984–1985, NAA: A6449, 2 (M–Z).

Pritchard, N, High Commissioner for the UK, letter to MC Timbs, Prime Minister's Department, 3 June 1960, NAA: A6456, R107/005.

Rickard, Doug, Royal Commission, 14–20 November 1984, NAA: A6448, 6.

Royal Commission into British Nuclear Tests in Australia, *Conclusions and Recommendations* (JR McClelland, J Fitch, WJA Jones), AGPS, Canberra, 1985.

Royal Commission into British Nuclear Tests in Australia, *Report* (JR McClelland, President), 2 vols, AGPS, Canberra, 1985.

Sherratt, Tim, 'A political inconvenience: Australian scientists at the British atomic weapons tests, 1952–53', *Historical Records of Australian Science*, vol. 6, no. 2, December 1985.

Stubbs, Stewart, Royal Commission, 5–30 May 1985, NAA: A6448, 14.

Sublette, Carey, 'Cobalt bombs and other salted bombs', Nuclear Weapon Archive, <nuclearweaponarchive.org/Nwfaq/Nfaq1.html#nfaq1.6>. Accessed June 2016.

Symonds, JL, *A History of British Atomic Tests in Australia*, AGPS, Canberra, 1985.

Titterton, EW, 'The race for atomic arms', *Sunday Herald*, 4 October 1953, p. 2.

Titterton, EW, *Facing the Atomic Future*, Macmillan, London, 1956.

Titterton, EW, Dwyer, LJ, Stevens, DJ & Keam, DW, 'Report on Measurement of Fall-Out Samples Collected between 15th May and 22nd July', 1957, tabled at AWTSC, 21st meeting, Department of Supply, Swanston St, Melbourne, 19 July 1957, NAA: A6456, R120/137 PART 1.

Titterton, EW, letter to JL Knott, Secretary, Department of Supply, 24 August 1960, NAA: 6456, R150/001.

Turner, Graeme, 'Of rocks and hard places: the colonized, the national and Australian cultural studies', *Cultural Studies*, vol. 6, no. 3, 1992.

7 INDIGENOUS PEOPLE AND THE BOMB TESTS

Assistant Secretary, Research and Development, Department of Supply, internal memorandum to Secretary, 2 May 1963, Malone files.

AWTSC, minutes of 21st meeting, Department of Supply, Swanston St, Melbourne, 19 July 1957, NAA: A6456, R120/137 PART 1.

Beale, Howard, Minister for Supply, 'Atomic Tests in Australia', top-secret Cabinet briefing document, submission no. 73, Canberra, 11 August 1954, Malone files.

Beale, Howard, letter to Paul Hasluck, Minister for Territories, 1 May 1957, Malone files.

Brown, HJ, Woomera, confidential cable re press reports about Aboriginal children being removed from their parents, to O'Connor, Department of Supply, 21 November 1956, Malone files.

Brown, HJ, Controller, Weapons Research Establishment, Salisbury, cable to EL Cook, Department of Supply, 2 December 1957, Malone files.

Butement, Alan, Chief Scientist, memorandum to Controller, Weapons Research Establishment, 8 June 1956, Malone files.

Dousset, Laurent, 'Politics and demography in a contact situation: the establishment of the Giles Meteorological Station in the Rawlinson Ranges, West Australia', *Aboriginal History*, vol. 26, 2002.

Duguid, Charles, 'The Central Aborigines Reserve', presidential address to Aborigines Advancement League of South Australia, 21 October 1957.

Expert Committee on the Review of Data on Atmospheric Fallout Arising from British Nuclear Tests in Australia, *Report* (CB Kerr, Chair) (Kerr Report), 31 May 1984, Malone files.

Foulkes, JN & Thompson, DS, 'A Biological Survey of the Maralinga Tjuratja Lands, South Australia, 2001–2007', Science Resource Centre, Information, Science and Technology Directorate, Department for Environment and Heritage, South Australia, December 2008.

Harman, JJ, Divisional Superintendent, Eastern Division, Western Australian Department of Native Welfare, cable to Secretary, Department of Supply, Melbourne, 1 May 1963, Malone files.

Kanytji (via interpreter), Royal Commission, 11 April – 2 May 1985, NAA: A6448, 13.

Kerr Report, see Expert Committee on the Review of Data on Atmospheric Fallout Arising from British Nuclear Tests in Australia.

Macaulay, R, 'Report on Mr. L. Beadell', letter to Superintendent, Woomera, 17 April 1957, Malone files.

Macaulay, R, 'Shell Lakes–Boundary Dam Patrol May–June 1959', letter to Commissioner, Native Welfare, Perth, Western Australia, 12 January 1960, Malone files.

Macaulay, R, 'Native Reserves of South-Western Central Australia – Their Future and Department of Supply Interest', report for Weapons Research Establishment, Woomera, 28 March 1960, Malone files.

Macaulay, R, report to Superintendent, Woomera, 16 October 1963, Malone files.

MacDougall, WB, reports to Superintendent, Woomera, 31 January 1952, 7 October 1953, 10 December 1953, July 1955, 11 October 1955, 16 January 1956, Malone files.

MacDougall, WB, 'Proposed Survey of Tribal Aborigines Living near the South Eastern Portion of the Central Reserve', sent to Superintendent, Woomera, 3 March 1952, Malone files.

MacDougall, WB, 'Detailed Survey of the Jangkuntjara Tribe – Their Traditional Tribal Country and Ceremonial Grounds', sent to Superintendent, Woomera, 2 March 1953, Malone files.

MacDougall, WB, 'Patrol at Ooldea', minute paper for Department of Supply, 5 February 1954, Malone files.

MacDougall, WB, letter to Secretary, Aborigines Protection Board, Adelaide, November 1954, Malone files.

MacDougall, WB, Native Patrol Officer, letter to Superintendent, Woomera, 4 November 1955, Malone files.

MacDougall, WB, 'Report on Patrol to Central Reserve Including That Area Recently Excised for the Use of the South West Mining Company', letter to Commissioner, Native Welfare, Western Australia, 3 August 1956, Malone files.

MacDougall, WB, 'Report on Patrol in Connection with Maralinga Tests 3rd September, to 13th October, 1957', report to Superintendent, Woomera, 18 October 1957, Malone files.

MacDougall, WB, 'Report on Patrol to Rawlinson Ranges 6th to 30th January, 1958', report to Superintendent, Woomera, 25 February 1958, Malone files.

Maralinga Peace Officer Guard, 'Native Movement – Maralinga', letter to Deputy Superintending Peace Officer, Adelaide, 16 May 1957, Malone files.

References

Mattingley, Christobel, 'Atom bombs before Aborigines: Maralinga', in Christobel Mattingley & Ken Hampton, *Survival in Our Own Land: 'Aboriginal' Experiences in 'South Australia' since 1836*, ALDAA in association with Hodder & Stoughton, Adelaide, 1988.

Mazel, Odette, 'Returning Parna Wiru: restitution of the Maralinga lands to traditional owners in South Australia', in Marcia Langton (ed.), *Settling with Indigenous People: Modern Treaty and Agreement Making*, Federation Press, 2006.

Milliken, Robert, *No Conceivable Injury*, Penguin, Ringwood, 1986.

Moroney, John, cable to HJ Brown, Controller, Weapons Research Establishment, Salisbury, 2 December 1957, Malone files.

O'Connor, FA, 'Atomic Tests in Australia – Report of Inter-Departmental Committee', briefing note for Minister for Supply, 29 September 1954, Malone files.

Penney, William, statement, Royal Commission, 1984–1985, NAA: A6449, 2 (M–Z).

Roach, WT, Meteorological Office, Bracknell & Ballis, DG, AWRE, 'Transport of Debris from the British Nuclear Test in South Australia on 15 October 1953', Royal Commission, 1984–1985, NAA: A6449, 2 (M–Z).

Royal Commission into British Nuclear Tests in Australia, *Conclusions and Recommendations* (JR McClelland, J Fitch, WJA Jones), AGPS, Canberra, 1985.

Royal Commission into British Nuclear Tests in Australia, *Report* (JR McClelland, President), 2 vols, AGPS, Canberra, 1985.

Secretary, Aborigines Protection Board, letters to Mr WB MacDougall, LRWE Range, Woomera, 15 December 1953 & 14 January 1955, Malone files.

Security Officer, Maralinga (name redacted), 'Control of Aborigines – Maralinga', letter to Range Commander, Maralinga, 25 July 1957, Malone files.

Smith, F, 'Report on Natives at PomPom – 14 May 57', letter to Harry Turner, Health Physics Adviser, 15 May 1957, Malone files.

Turner, OH, 'Health Physics Report on Natives at Pom Pom – 14 May 57', letter to Range Commander, Maralinga, 18 May 1957, Malone files.

Williams, Geoff & O'Brien, Richard, appraisal of *The Black Mist and Its Aftermath – Oral Histories by Lallie Lennon*, ARPANSA, Melbourne, 20 April 2010.

Wills, HA, Chief Executive Officer, Maralinga Committee, 'Aboriginal Population of South Australia', letter to HJ Brown, Controller, Weapons Research Establishment, Salisbury, 6 June 1956, Malone files.

Wright, RVS, 'Bates, Daisy May (1863–1951)', *Australian Dictionary of Biography*, vol. 7, Melbourne University Press, Melbourne, 1979.

Yalata and Oak Valley Communities with Mattingley, Christobel, *Maralinga: The Anangu Story*, Allen & Unwin, Sydney, 2009.

8 D-NOTICES AND MEDIA SELF-CENSORSHIP

Admiralty, War Office, Air Ministry and Press Committee, Department of Defence confidential briefing document, NAA: A5954, 1956/6.

Buchanan, AE, '"D" Notice no. 8 – Atomic Tests', [1952], NAA: A816, 10/301/131.

Buchanan, AE, '"D" Notices – Press Agencies', minute paper for Head of Department of Defence Frederick Shedden, 2 September 1952, NAA: A816, 10/301/132.

Buchanan, AE, 'Atomic Test – Official Releases', letter to members of Defence, Press and Broadcasting Committee, 19 September 1952, NAA: A816, 10/301/129.

Buchanan, AE, '"D" Notice no. 8 Atomic Tests – Cancellation', 10 November 1952, NAA: A816, 10/301/132.

Buchanan, AE, '"D" Notice no. 10', Defence, Press and Broadcasting Committee, [1953], NAA: A1209, 1957/5486.

Commission of Enquiry into the Australian Secret Intelligence Service, 'The D-notice system', in *Report on the Secret Intelligence Service*, public edn, AGPS, Canberra, March 1995, Chapter 11.

Cook, EL, Department of Supply, secret minute paper containing briefing on D-notices, 11 December 1951, NAA: A1209, 1957/5486.

Cryle, Denis, 'Rousing the British-speaking world: Australian newspaper proprietors and freedom of the press, 1940–1950', ARC-funded project, Central Queensland University, paper at Australian Media Traditions Conference, 2007.

Curry Jansen, Sue & Martin, Brian, 'Exposing and opposing censorship: backfire dynamics in freedom-of-speech struggles', *Pacific Journalism Review*, vol. 10, no. 1, April 2004, pp. 29–45.

de Burgh, EC, Editor, *West Australian*, letter to AE Buchanan, 5 August 1952, NAA: A816, 10/301/129.

Defence, Press and Broadcasting Committee, minutes of 1st meeting, 14 July 1952, NAA: A1209, 1957/5486.

Doutreband, R, letter to Robert Menzies, 5 December 1950, NAA: A816, 10/301/128.

Fadden, Arthur, 'UK Atomic Test Woomera – D-notice to the Press', letter to Press Chiefs, 26 June 1953, in JL Symonds, *A History of British Atomic Tests in Australia*, AGPS, Canberra, 1985, p. 151.

Fairley, Douglas, 'D notices, official secrets and the law', *Oxford Journal of Legal Studies*, vol. 10, no. 3, Autumn 1990.

General Manager, John Fairfax & Sons Pty Ltd, letter to Prime Minister Robert Menzies, 19 August 1952, NAA: A816, 10/301/129.

Macintyre, Stuart, *A Concise History of Australia*, 2nd edn, Cambridge University Press, Port Melbourne, 2004.

Maher, Laurence W, 'H.V. Evatt and the Petrov defection: a lawyer's interpretation', *Australian Society of Labour History*, 2007, <www.historycooperative.org/proceedings/asslh2/maher.html>. Accessed 31 December 2009.

Martin, AW, *Robert Menzies: A Life*, vol. 2, *1944–1978*, Melbourne University Press, Melbourne, 1999.

McBride, PA, letter to Prime Minister Robert Menzies, 28 March 1952, NAA: A5954, 1956/6.

McBride, PA, letter to Joseph Francis, William McMahon, EJ Harrison & Howard Beale, 2 July 1952, NAA: A5954, 1956/6.

McBride, PA, letter to Robert Menzies, 26 August 1952, NAA: A816, 10/301/129.

McCadden, GE, United Press Associations, letter to Michael Byrne, 17 June 1952, NAA: A816, 10/301/129.

Menzies, Robert, 'Security of Defence Information', letter to E Kennedy, President, Australian Newspaper Proprietors' Association, 22 November 1950, NAA: A5954, 1956/6.

Menzies, Robert, letter to Eric Kennedy, Chief Executive Officer, Associated Newspapers, 15 August 1952, NAA: A816, 10/301/129.

Menzies, Robert, *The Measure of the Years*, Cassell Australia, Melbourne, 1970.

Milliken, Robert, *No Conceivable Injury*, Penguin, Ringwood, 1986.

O'Connor, Frank, Department of Supply, letter to Major General EL Sheehan, Australian Defence Representative, London, 14 July 1953, NAA: A5954, 1594/2.

O'Connor, FA, Department of Supply, letter to Secretary, Prime Minister's Department, 20 July 1955, NAA: A1209, 1957/5486.

Office of the Secretary of Defense, memorandum for the press, 29 March 1948, NAA: A816, 10/301/130.

Packer, Frank, letter to Robert Menzies, 4 December 1950, NAA: A816, 10/301/129.

Resolution adopted by representatives of US media, 29 March 1948, NAA: A816, 10/301/130.

Sadler, Pauline, 'The D-notice system', *Press Council News*, vol. 12, no. 2, May 2000, <www.presscouncil.org.au/pcsite/apcnews/may00/dnote.html>.

Sadler, Pauline, *National Security and the D-notice System*, Dartmouth Publishing Company, Aldershot, 2001.

Schedvin, CB, *Shaping Science and Industry: A History of Australia's Council for Scientific and Industrial Research, 1926–49*, Allen & Unwin, Sydney, 1987.

Shedden, FG, letter to JT Pinner, Public Service Board of Commissioners, 2 July 1952, NAA: A816, 10/301/128.

Shedden, FG, letter to ASJ Brown, Prime Minister's Department, 16 July 1952, NAA: A816, 10/301/129.

Sydney Morning Herald editorial, 18 July 1940, reproduced in FK Crowley, *Modern Australian Documents 1939–1970*, vol. 2, Wren Publishing, Melbourne, 1973.

Thomson, George, Admiralty, War Office, Air Ministry and Press Committee, letter to Major General EL Sheehan, Australian Joint Services Staff, London, 17 June 1953, NAA: A5954, 1594/2.

Thomson, George, Secretary, Admiralty, War Office, Air Ministry and Press Committee (UK), 'Summary of "D" Notices in Force and of Private and Confidential Letters Issued up to 1st May, 1954', 17 May 1954, NAA: A816, 10/301/130.

'U.S. alarm at leakage of Defence secrets', *Herald* (Melbourne), 8 March 1948, clippings package, NAA: A5954, 1956/6.

9 CLEAN-UPS AND COVER-UPS

Arnold, Lorna & Smith, Mark, *Britain, Australia and the Bomb: The Nuclear Tests and Their Aftermath*, 2nd edn, Palgrave Macmillan, New York, 2006.

Borshmann, Gregg, 'Maralinga: the fall out continues', *Background Briefing*, ABC Radio National, 16 April 2000.

Burns, Peter & Williams, Geoff, interview with author, ARPANSA, Melbourne, 15 April 2004.

Department of Education, Science and Training, 'Rehabilitation of Former Nuclear Tests Sites at Emu and Maralinga (Australia) 2003' (MARTAC Report), Commonwealth of Australia, Canberra, 2002.

MARTAC Report, see Department of Education, Science and Training.

Moroney, John, 'Statement on the Disagreement in the Plutonium Data at Maralinga in the Discussion with the Australian Editor of *New Scientist*', briefing for Geoff Williams & Pat Davoren, 3 June 1993.

'Noah Pearce', Geni, <www.geni.com/people/Noah-Pearce/6000000000836708056>. Accessed 1 November 2015.

Owen, George, statement, Royal Commission, 1984–1985, NAA: A6449, 2 (M–Z).

Owen, George, Royal Commission, 3–30 January 1985, NAA: A6448, 9.

Parkinson, Alan, *Maralinga: Australia's Nuclear Waste Cover-Up*, ABC Books, Sydney, 2007.

Pearce, Noah, 'Final Report on Residual Radioactive Contamination of the Maralinga

Range and the Emu Site', AWRE report no. 01–16/68 (Pearce Report), UK Atomic Energy Authority, January 1968. Edited version released by Australian Department of National Development, May 1979. Full version tabled by Australian Senate, May 1984.

Pearce, Noah, statement, Royal Commission, 1984–1985, NAA: A6449, 2 (M–Z).

Pearce, Noah, Royal Commission, 4–18 March 1985, NAA: A6448, 12.

Pearce Report, see Pearce, Noah, 'Final Report on Residual Radioactive Contamination of the Maralinga Range and the Emu Site'.

Richardson, JF, Australian Health Physics Representative, 'Report to the Australian Atomic Weapons Tests Safety Committee on Visits to Maralinga during Operation Brumbie [sic]', 19 July 1967, Malone files.

Royal Commission into British Nuclear Tests in Australia, Report (JR McClelland, President), 2 vols, AGPS, Canberra, 1985.

Sherratt, Tim, 'A political inconvenience: Australian scientists at the British atomic weapons tests, 1952–53', Historical Records of Australian Science, vol. 6, no. 2, December 1985.

Symonds, JL, A History of British Atomic Tests in Australia, AGPS, Canberra, 1985.

10 MEDIA, POLITICS AND THE ROYAL COMMISSION

'After Maralinga' (editorial), Advertiser, 17 April 1980, p. 5.

Australia, House of Representatives, Debates, vol. 102, 1976, p. 3574.

Australia, House of Representatives, Debates, vol. 104, 1977, p. 880.

Australia, House of Representatives, Debates, 'Radioactive Material at Maralinga', ministerial statement by Mr Killen, 11 October 1978.

Baker, Candida, 'A new eye focuses on the rich and powerful', The Age, 15 August 1987, p. 3.

Ball, Robert, 'A "devil spirit" that didn't go', Advertiser, 3 May 1980, p. 1.

Bayly, Brett, 'Killen orders enquiry into Maralinga', Advertiser, 10 December 1976, p. 1.

Beale, Howard, quoted in 'Plutonium scare "ridiculous"', SMH, 11 October 1978, p. 2.

Blakeway, Denys & Lloyd-Roberts, Sue, Fields of Thunder: Testing Britain's Bomb, George Allen & Unwin, London, 1985.

Bowers, Peter, 'Maralinga's plutonium mystery becomes even deeper: who left nuclear lump at the site?', SMH, 7 October 1978, p. 4.

Bowers, Peter, 'British team to check atom dump', SMH, 11 October 1978, p. 1.

Bowers, Peter, 'The real mystery of Maralinga', SMH, 16 October 1978, p. 7.

Cadogan-Cowper, GF, Senior Adviser, Resources Branch, briefing paper for Prime Minister Malcolm Fraser, 13 October 1978, NAA: A6456, R065/080.

Carrick, Senator John, draft reply to Senator Evans' question, 16 November 1978, NAA: A6456, R188/014.

Cass, Moss, Minister for the Environment and Conservation, letter to Lance Barnard, Minister for Defence, 3 December 1974, NAA: A6456, R065/080.

Clark, Emeritus Professor AM, Chair, AIRAC, letter to Barry Cohen, Minister for Home Affairs and Environment, 13 August 1984, Malone files.

Colless, Malcolm, 'Govt acts over N-waste safety: Killen loses nuclear role', The Australian, 10 November 1978, p. 1.

Colless, Malcolm, '"Take it back" requests ignored: British snub on plutonium plea', The Australian, 16 November 1978, p. 1.

Connor, Steve, 'WA atom blast was far bigger, UK says', SMH, 26 May 1984, p. 3.

References

Cook, Melbourne, cable on press arrangements to Herington, Adelaide, 11 September 1956, NAA: A6456, R047/011.

Dean, Anabel, 'First win for Maralinga victim and hope for others', *SMH*, 23 December 1988, p. 3.

Defence Reporter, 'Doubts cast on SA nuclear dump', *SMH*, 11 February 1977, p. 3.

Department of Foreign Affairs, Canberra, outward cablegram to London, 14 & 19 November 1978, NAA: A6456, R188/014.

Dudgeon, Henry, confidential cable to Foreign and Commonwealth Office, 3 November 1978, National Archives of the UK, FCO 1/20.

Dunstan, Don, letter to Malcolm Fraser, tabled as evidence to Senate Estimates Committee, 17 October 1978, NAA: A6456, R188/014.

English, David & De Ionno, Peter, 'New claims of A-test link with cancer', *Advertiser*, 17 April 1980, p. 1.

English, David & De Ionno, Peter, 'SA atom tests: was cost too high?', *Advertiser*, 18 April 1980, p. 9.

Evans, Senator Gareth, draft question for Senate, 16 November 1978, NAA: A6456, R188/014.

Fernandez, RR, Acting Deputy Secretary, Department of Foreign Affairs, teletype message to Minister for Foreign Affairs Andrew Peacock, 14 December 1976, NAA: A6456, R065/080.

Foreign Affairs and Defence Committee, Cabinet minute, decision no. 6812 (FAD), Canberra, 28 September 1978, NAA: A12909, 2605.

Fraser, Malcolm, Prime Minister of Australia, confidential teletype message to Don Dunstan, Premier of South Australia, 20 October 1978, NAA: A6456, R188/014.

Fraser, Malcolm, Prime Minister of Australia, letter to South Australian Premier Don Dunstan, 12 February 1979, Malone files.

Garland, RV, Minister for Supply, letter to P Howson, Minister for the Environment, Aborigines and the Arts, Canberra, 2 February 1972, Malone files.

Garland, RV, extract from Australia, House of Representatives, *Debates*, 14 September 1972, attached to GF Cadogan-Cowper, Senior Adviser, Resources Branch, briefing paper for Prime Minister Malcolm Fraser, 13 October 1978, NAA: A6456, R065/080.

Hailstone, Barry, 'Expert calls for enquiry into radioactive waste', *Advertiser*, 9 December 1976, p. 1.

Jory, Rex, '3 government reports now confirm ... plutonium buried at Maralinga', *The News*, 17 December 1976, p. 1.

Kelso, JR, Acting First Assistant Secretary, Nuclear Affairs Division, Foreign Affairs ministerial submission on Maralinga, 15 November 1978, NAA: A6456, R188/014.

'Killen attacks Review report', *SMH*, 12 October 1978, p. 9.

Killen, DJ, Minister for Defence, 'Plutonium Buried near Maralinga Airfield', Cabinet submission no. 2606, 11 September 1978, NAA: A12909, 2605.

Kruger, Andrew, 'Britain asked to take atomic waste', *SMH*, 10 October 1978, p. 1.

Kruger, Andrew, 'Full story not told on waste – Hayden', *SMH*, 13 October 1978, p. 13.

MacCallum, Mungo, 'Labor's Maralinga attack very non-nuclear', *Australian Financial Review*, 11 October 1978, p. 4.

MacCallum, Mungo, 'Killen throws a Maralinga bomb – with fallout', *Australian Financial Review*, 12 October 1978, p. 4.

MacCallum, Mungo, 'Toohey: making red faces in high places', *West Australian*, 2 September 1989, p. 9.

Macintyre, Stuart, *A Concise History of Australia*, 2nd edn, Cambridge University Press, Port Melbourne, 2004.

Malone, Paul & Conkey, Howard, 'Confusion on contamination: radiation too high at Maralinga site', *Canberra Times*, 30 September 1984, p. 12.

McClelland, James, transcript of interview with Robin Hughes, Australian Biography Project, 25 January 1995, <www.australianbiography.gov.au/subjects/mcclelland/interview6.html>. Accessed 20 November 2015.

Michel, Dieter, 'Villains, victims and heroes: contested memory and the British nuclear tests in Australia', *Journal of Australian Studies*, no. 80, 2004.

Milliken, Robert, *No Conceivable Injury*, Penguin, Ringwood, 1986.

Morgan, Sue, 'Britain's star witness sheds little light', *SMH*, 2 February 1985, p. 12.

Morgan, Sue, 'Titterton was "a British plant"', *SMH*, 6 December 1985, p. 1.

Morgan, Sue & Casey, Andrew, 'Atomic tests report urges compensation and multi-million-dollar clean-up: Thatcher set to tough it out', *SMH*, 6 December 1985, p. 1.

Norris, Robert Standish, *Questions on the British H-Bomb*, Natural Resources Defense Council, 22 June 1992.

'Nuclear dump "does exist" at Maralinga', *Advertiser*, 4 December 1976, p. 3.

'Nuclear waste could be stolen, govt told', *SMH*, 5 October 1978, p. 1.

'Nuclear waste dump in SA: ex-RAAF man', *Advertiser*, 3 December 1976, p. 1.

Pincher, Chapman, quoted in 'Views from the past', sidebar story attached to David English & Peter De Ionno, 'Fall-out blankets a sleeping city', *Advertiser*, 17 April 1980, p. 10.

'Plutonium "unlikely"', sidebar to Andrew Kruger, 'Britain asked to take atomic waste', *SMH*, 10 October 1978, p. 1.

Royal Commission into British Nuclear Tests in Australia, *Report* (JR McClelland, President), 2 vols, AGPS, Canberra, 1985.

Shelton, JP, Acting Director, Department of Environment, minute paper, 14 December 1976, NAA: A6456, R065/080.

Summers, Anne, *Gamble for Power: How Bob Hawke Beat Malcolm Fraser; The 1983 Federal Election*, Thomas Nelson Australia, Melbourne, 1983.

Symonds, JL, 'British Atomic Tests in Australia Chronology of Events: 1950–1968', 1984–1985, NAA: A6456/3, R023/003.

Symonds, JL, *A History of British Atomic Tests in Australia*, AGPS, Canberra, 1985.

Symons, Michael, 'Is it Russian roulette?', *SMH*, 16 June 1972, p. 7.

Tame, Adrian & Robotham, FP, *Maralinga: British A-Bomb Australian Legacy*, Dominion Press, Melbourne, 1982.

Titterton, EW, 'The Maralinga scare', *SMH*, 7 October 1978, p. 10.

Toohey, Brian, 'Killen warns on plutonium pile', *Australian Financial Review*, 5 October 1978, pp. 1, 6.

Toohey, Brian, 'Maralinga: the "do nothing" solution', *Australian Financial Review*, 11 October 1978, pp. 1, 10, 37.

Toohey, Brian, 'Govt may exhume plutonium waste', *Australian Financial Review*, 12 October 1978, pp. 1, 10, 14.

Toohey, Brian, 'Maralinga issue raises Defence Dept question', *Australian Financial Review*, 13 October 1978, pp. 7, 12.

Toohey, Brian, 'Plutonium on the wind: The terrible legacy of Maralinga', *National Times*, 4–10 May 1984, pp. 3–5.

Toohey, Brian, personal correspondence with author, 31 December 2009.

Toohey, Brian & Wilkinson, Marian, *The Book of Leaks: Exposés in Defence of the Public's Right to Know*, Angus and Robertson, North Ryde, 1987.

Uren, Tom, *Straight Left*, Vintage, Milsons Point, 1995.

'The usual four guard the plutonium', *SMH*, 7 October 1978, p. 1.

Walsh, Peter, statement, Senate, 4 May 1984, reproduced in *Australian Foreign Affairs Record*, vol. 55, no. 5, May 1984, p. 486.

Walsh, Peter, media release, 15 May 1984, reproduced in *Australian Foreign Affairs Record*, vol. 55, no. 5, May 1984, p. 547.

Walsh, Peter, statement, Senate, 7 June 1984, reproduced in *Australian Foreign Affairs Record*, vol. 55, no. 6, June 1984, p. 636.

Walsh, Peter, media release, 5 July 1984, reproduced in *Australian Foreign Affairs Record*, vol. 55, no. 7, July 1984, p. 732.

Walsh, Peter, *Confessions of a Failed Finance Minister*, Random House Australia, Milsons Point, 1995.

Williams, Geoff, ARPANSA, personal correspondence with author, 27 July 2010.

Woodward, Lindy, 'Buffalo Bill and the Maralingerers', *New Journalist*, no. 43, 1984, p. 18.

11 THE ROLLER COASTER INVESTIGATION

ABC News South Australia, 12 June 1993, various bulletins.

Anderson, Ian, 'Australia counts the cost of Maralinga cleanup', *New Scientist*, 17 November 1990.

Anderson, Ian, interview with unknown interviewee (ministerial staffer), 1993 (exact date unknown), Anderson files.

Anderson, Ian, interview on *Daybreak Show*, June 1993, Anderson files.

Anderson, Ian, interview on Tony Delroy's late night show, ABC Radio National, 10 June 1993, Anderson files.

Anderson, Ian, 'Britain's dirty deeds at Maralinga', *New Scientist*, 12 June 1993, p. 13.

Anderson, Ian, commentary on *The Science Show* (host Robyn Williams), ABC Radio National, 12 June 1993, Anderson files.

Anderson, Ian, statement in support of application for a Michael Daley award, 29 September 1993, Anderson files.

Armitage, Catherine, 'Dead zone', *SMH*, 10 June 1993.

Arnold, Lorna & Smith, Mark, *Britain, Australia and the Bomb: The Nuclear Tests and Their Aftermath*, 2nd edn, Palgrave Macmillan, New York, 2006.

Burns, Peter & Williams, Geoff, tape recording of interview with Ian Anderson, 1993 (exact date unknown), Anderson files.

Burns, Peter & Williams, Geoff, interview with author, ARPANSA, Melbourne, 15 April 2004.

Davoren, Patrick, personal correspondence with author, 13 May 2004.

Department of Education, Science and Training, 'Rehabilitation of Former Nuclear Tests Sites at Emu and Maralinga (Australia) 2003' (MARTAC Report), Commonwealth of Australia, Canberra, 2002.

Hamilton, Archie, House of Commons, *Debates*, vol. 222, 1 April 1993, cc. 650–651, <hansard.millbanksystems.com/commons/1993/apr/01/atomic-test-site-south-australia#column_650>. Accessed 23 March 2016.

John, Daniel & Black, Ian, 'Nuclear clean-up settlement near', *Guardian*, 18 June 1993, p. 14.

Jones, Philip, 'Ian Anderson: journalist who exposed Britain's dirty nuclear deeds in Australia', *Guardian*, 5 April 2000.

MARTAC Report, see Department of Education, Science and Training.

Moroney, John, letter to Patrick Davoren, 28 November 1991.

Moroney, John, 'Aide Memoire on the Nuclear Tests in Australia in the Context of the British Weapons Development Program', unpublished briefing document, ARL, 24 December 1992.

Moroney, John, annotated draft of Ian Anderson's *New Scientist* story, unpublished, 2 June 1993, Anderson files.

Moroney, John, 'Statement on the Disagreement in the Plutonium Data at Maralinga in the Discussion with the Australian Editor of *New Scientist*', briefing for Geoff Williams & Pat Davoren, 3 June 1993.

New Scientist editorial comment, 12 June 1993, p. 3.

O'Shea, Darcy, 'Maralinga: righting the wrongs', *Habitat Australia*, vol. 19, no. 3, June 1991, p. 22.

'Rehabilitation of the former nuclear tests sites at Maralinga and Emu', in *Annual Report of the Chief Executive Officer of ARPANSA, 1998–1999*, ARPANSA, Melbourne, 1999, pp. 35–38.

Royal Commission into British Nuclear Tests in Australia, *Report* (JR McClelland, President), 2 vols, AGPS, Canberra, 1985.

The Science Show (host Robyn Williams), ABC Radio National, 29 November 1997.

Stenberg, Maryann, 'Tests show a wider Maralinga cover-up', *The Age*, 11 June 1993.

Thorn, Robert N & Westervelt, Donald R, 'Hydronuclear Experiments', report for Los Alamos National Laboratories, LA-10902-MS, February 1987.

Thwaites, Tim, 'Ian Anderson: the end of an era', *New Scientist*, 25 March 2000.

Webb, Jeremy, *New Scientist*, personal correspondence with author, 25 May 2004.

12 THE REMAINS OF MARALINGA

Arvanitakis, James, 'Staging Maralinga and looking for community (or why we must desire community before we can find it)', *Research in Drama Education*, vol. 13, no. 3, pp. 295–306.

AWTSC, minutes of 19th meeting, Commonwealth X-Ray and Radium Laboratory, Melbourne, 11 June 1957, NAA: A6456, R120/137 PART 1.

Barnaby, Frank, 'The management of radioactive wastes and the disposal of plutonium', paper at Medical Association for Prevention of War Conference, 2000.

Brown, Paul, 'British nuclear testing in Australia: performing the Maralinga experiment through verbatim theatre', *Journal and Proceedings of the Royal Society of New South Wales*, vol. 139, 2006.

Clarke QC, John, 'British atomic tests', in Department of Veteran Affairs, *Review of Veterans' Entitlements*, Canberra, 6 January 2003, Chapter 16.

Cross, Roger & Hudson, Avon, *Beyond Belief: The British Bomb Tests; Australia's Veterans Speak Out*, Wakefield Press, Kent Town, 2005.

Department of Education, Science and Training, 'Rehabilitation of Former Nuclear Tests Sites at Emu and Maralinga (Australia) 2003' (MARTAC Report), Commonwealth of Australia, Canberra, 2002.

Gellman, B, 'Edward Snowden, after months of NSA revelations, says his mission's Aaccomplished', *Washington Post*, 23 December 2013.

James, Colin & Maiden, Samantha, 'Secret atomic child files opened', *Advertiser*, 5 September 2001.

Jean, Peter, 'Aborigines now control ex-nuke site for tourism', *Advertiser*, 4 June 2014.

References

Keane, John, 'Maralinga's afterlife', *The Age*, 11 May 2003.

'Maralinga lament', Australian Nexus, <www.ausnexus.co.uk/events/maralinga
-lament/#.VlERyUYXgQ0>. Accessed 22 November 2015.

Maralinga Tours, <www.maralingatours.com.au>. Accessed 22 November 2015.

MARTAC Report, see Department of Education, Science and Training.

Moroney, John, letter to Ernest Titterton, 19 August 1966, NAA: A6456, R120/111.

Moroney, John, letter to WJ Gibbs, Bureau of Meteorology, 26 August 1966, NAA:
A6456, R120/111.

Nuclear Futures, <nuclearfutures.org>. Accessed 22 November 2015.

Parkinson, Alan, 'The Maralinga rehabilitation project: final report', *Journal of Medicine,
Conflict and Survival*, vol. 20, 2004.

Pearce, Noah, letter to John Moroney, 9 August 1966, NAA: A6456, R120/111.

Rankin, Scott, 'Ngapartji Ngapartji', in *Namatjira / Ngapartji Ngapartji*, Currency Press,
Sydney, 2012.

Rankin, Scott, 'Ngapartji Ngapartji', Wikipedia, <en.wikipedia.org/wiki/Ngapartji_
Ngapartji>. Accessed 26 March 2016.

Spinardi, Graham, 'Aldermaston and British nuclear weapons development: testing the
"Zuckerman thesis"', *Social Studies of Science*, vol. 27, no. 4, 1997, pp. 547–582.

Titterton, EW, letter to John Moroney, 23 August 1966, NAA: A6456, R120/111.

Tooke, Thomas Frederick, Royal Commission, 31 October – 13 November 1984, NAA:
A6448, 5.

Tooke, Thomas Frederick, statement, Royal Commission, 1984–1985, NAA: A6450, 5
(T–Z).

Walker, Frank, *Maralinga: The Chilling Exposé of Our Secret Nuclear Shame and Betrayal of
Our Troops and Country*, Hachette Australia, Sydney, 2014.

WikiLeaks, 'How Britain got the bomb: Britain's "Oppenheimer", William G. Penney
and his secret report on what Britain needed to do to "get the bomb"',
<wikileaks.org/wiki/How_Britain_got_the_bomb#Penney_and_the_Start_of_
the_Post-War_British_Atomic_Bomb_Program>. Accessed 19 November 2015.

Woollett, SM, 'Rehabilitating Maralinga', *Clean Air and Environmental Quality*, vol. 37,
issue 4, November 2003.

APPENDIX

Royal Commission into British Nuclear Tests in Australia, *Report* (JR McClelland,
President), vol. 2, AGPS, Canberra, 1985, Appendix G, Chronology, pp. vii–1 to
vii–10.

GLOSSARY

NTD Resource Center, 'Stochastic effects', <www.nde-ed.org/EducationResources/
CommunityCollege/RadiationSafety/biological/stochastic/stochastic.htm>.
Accessed 27 March 2016.

Royal Commission into British Nuclear Tests in Australia, *Report* (JR McClelland,
President), vol. 2, AGPS, Canberra, 1985, pp. II–1 to II–9.

Symonds, JL, *A History of British Atomic Tests in Australia*, AGPS, Canberra, 1985,
pp. 585–593.

Bibliography

A note about sources

Most of my primary sources have come from the National Archives of Australia (NAA), the National Library of Australia and the National Archives of the United Kingdom (Kew). However, I have also been fortunate to have access to other equally crucial forms of primary material as well. The late *New Scientist* journalist Ian Anderson kept extensive files, including audio recordings. I was granted access to some of these early in my PhD study, and have used several of them in this book. In addition, Dr Geoff Williams at ARPANSA gave me access to previously unpublished material from his late colleague John Moroney, particularly related to the Roller Coaster investigation and his side of the 1993 *New Scientist* story. Also, the former *Canberra Times* journalist Paul Malone received numerous files from government departments, notably the Department of Defence and the Department of Supply, under an accelerated document release program in the 1980s. Paul generously gave me access to these files, which do not have NAA numbers. The same files may well have found their way to the NAA, but I have used the unnumbered files provided by Paul. I note these particular files in the relevant parts of the bibliographical details.

MEDIA REPORTS
Authors named
Anderson, Ian, 'Australia counts the cost of Maralinga cleanup', *New Scientist*,
17 November 1990.
Anderson, Ian, 'Britain's dirty deeds at Maralinga', *New Scientist*, 12 June 1993, pp. 12–13.
Anderson, Ian, commentary on *The Science Show* (host Robyn Williams), ABC Radio
National, 12 June 1993, Anderson files.
Armitage, Catherine, 'Dead zone', *Sydney Morning Herald* (*SMH*), 10 June 1993.
Baker, Candida, 'A new eye focuses on the rich and powerful', *The Age*, 15 August
1987, p. 3.

Ball, Robert, 'A "devil spirit" that didn't go', *Advertiser*, 3 May 1980, p. 1.

Bayly, Brett, 'Killen orders enquiry into Maralinga', *Advertiser*, 10 December 1976, p. 1.

Beale, Howard, 'Why Australia provides the site for atomic tests', *SMH*, 13 August 1956, p. 2.

Beale, Howard, quoted in 'Plutonium scare "ridiculous"', *SMH*, 11 October 1978, p. 2.

Borshmann, Gregg, 'Maralinga: the fall out continues', *Background Briefing*, ABC Radio National, 16 April 2000.

Bowers, Peter, 'Maralinga's plutonium mystery becomes even deeper: who left nuclear lump at the site?', *SMH*, 7 October 1978, p. 4.

Bowers, Peter, 'British team to check atom dump', *SMH*, 11 October 1978, p. 1.

Bowers, Peter, 'The real mystery of Maralinga', *SMH*, 16 October 1978, p. 7.

Colless, Malcolm, 'Govt acts over N-waste safety: Killen loses nuclear role', *The Australian*, 10 November 1978, p. 1.

Colless, Malcolm, '"Take it back" requests ignored: British snub on plutonium plea', *The Australian*, 16 November 1978, p. 1.

Connor, Steve, 'WA atom blast was far bigger, UK says', *SMH*, 26 May 1984, p. 3.

Dean, Anabel, 'First win for Maralinga victim and hope for others', *SMH*, 23 December 1988, p. 3.

English, David & De Ionno, Peter, 'New claims of A-test link with cancer', *Advertiser*, 17 April 1980, p. 1.

English, David & De Ionno, Peter, 'SA atom tests: was cost too high?', *Advertiser*, 18 April 1980, p. 9.

Gellman, B, 'Edward Snowden, after months of NSA revelations, says his mission's accomplished', *Washington Post*, 23 December 2013.

Hailstone, Barry, 'Expert calls for enquiry into radioactive waste', *Advertiser*, 9 December 1976, p. 1.

James, Colin & Maiden, Samantha, 'Secret atomic child files opened', *Advertiser*, 5 September 2001.

Jean, Peter, 'Aborigines now control ex-nuke site for tourism', *Advertiser*, 4 June 2014.

John, Daniel & Black, Ian, 'Nuclear clean-up settlement near', *Guardian*, 18 June 1993, p. 14.

Jones, Philip, 'Ian Anderson: journalist who exposed Britain's dirty nuclear deeds in Australia', *Guardian*, 5 April 2000.

Jory, Rex, '3 government reports now confirm ... plutonium buried at Maralinga', *The News*, 17 December 1976, p. 1.

Keane, John, 'Maralinga's afterlife', *The Age*, 11 May 2003.

Kruger, Andrew, 'Britain asked to take atomic waste', *SMH*, 10 October 1978, p. 1.

Kruger, Andrew, 'Full story not told on waste – Hayden', *SMH*, 13 October 1978, p. 13.

MacCallum, Mungo, 'Labor's Maralinga attack very non-nuclear', *Australian Financial Review*, 11 October 1978, p. 4.

MacCallum, Mungo, 'Killen throws a Maralinga bomb – with fallout', *Australian Financial Review*, 12 October 1978, p. 4.

MacCallum, Mungo, 'Toohey: making red faces in high places', *West Australian*, 2 September 1989, p. 9.

Malone, Paul & Conkey, Howard, 'Confusion on contamination: radiation too high at Maralinga site', *Canberra Times*, 30 September 1984, p. 12.

Moore, Daniel, '30 years on: Britain agrees to pay its Maralinga victims', *SMH*, 4 April 1988, p. 1.

Morgan, Sue, 'Britain's star witness sheds little light', *SMH*, 2 February 1985, p. 12.

Morgan, Sue, 'Titterton was "a British plant"', *SMH*, 6 December 1985, p. 1.

Morgan, Sue & Casey, Andrew, 'Atomic tests report urges compensation and multi-million-dollar clean-up: Thatcher set to tough it out', *SMH*, 6 December 1985, p. 1.

Nason, David, 'Owners to reclaim Maralinga bomb site', *The Australian*, 10 November 2009, p. 7.

Ophel, TR, 'Sir Ernest William Titterton', Obituary, *ANU Reporter*, 23 February 1990, p. 4.

O'Shea, Darcy, 'Maralinga: righting the wrongs', *Habitat Australia*, vol. 19, no. 3, June 1991, p. 22.

Pincher, Chapman, 'Could be dropped by R.A.F. plane', *SMH*, 16 October 1953, p. 1.

Pincher, Chapman, quoted in 'Views from the past', sidebar story attached to David English & Peter De Ionno, 'Fall-out blankets a sleeping city', *Advertiser*, 17 April 1980, p. 10.

Porter, David, 'Atomic bomb lessons in the desert', *SMH*, 22 June 1988, p. 2.

The Science Show (host Robyn Williams), ABC Radio National, 29 November 1997.

Stanton, John, 'Plutonium dumps a risk "for thousands of years"', *The Australian*, 25 May 1984, p. 2.

Stenberg, Maryann, 'Tests show a wider Maralinga cover-up', *The Age*, 11 June 1993.

Symons, Michael, 'Is it Russian roulette?', *SMH*, 16 June 1972, p. 7.

Thwaites, Tim, 'Ian Anderson: the end of an era', *New Scientist*, 25 March 2000.

Titterton, EW, 'The race for atomic arms', *Sunday Herald*, 4 October 1953, p. 2.

Titterton, EW, 'Answers to questions on atomic tests', *The Age*, 15 May 1956, p. 2.

Titterton, EW, 'Some questions and answers on latest atom tests', *SMH*, 15 May 1956, p. 2.

Titterton, EW, 'Why Australia is atom testing ground', *The Age*, 16 May 1956, p. 2.

Titterton, EW, 'Why Australia is preferred as an atom testing ground', *SMH*, 16 May 1956, p. 2.

Titterton, EW, 'After atomic bomb tests ... how dangerous is the mushroom cloud?', *The Age*, 19 July 1956, p. 2.

Titterton, EW, 'The Maralinga scare', *SMH*, 7 October 1978, p. 10.

Toohey, Brian, 'Killen warns on plutonium pile', *Australian Financial Review*, 5 October 1978.

Toohey, Brian, 'Maralinga: the "do nothing" solution', *Australian Financial Review*, 11 October 1978, pp. 1, 10, 37.

Toohey, Brian, 'Govt may exhume plutonium waste', *Australian Financial Review*, 12 October 1978, pp. 1, 10, 14.

Toohey, Brian, 'Maralinga issue raises Defence Dept question', *Australian Financial Review*, 13 October 1978, pp. 7, 12.

Toohey, Brian, 'Plutonium on the wind: the terrible legacy of Maralinga', *National Times*, 4–10 May 1984, pp. 3–5.

Woodward, Lindy, 'Buffalo Bill and the Maralingerers', *New Journalist*, no. 43, April 1984, p. 18.

Authors unnamed

(chronological order)

Sydney Morning Herald editorial, 18 July 1940, reproduced in FK Crowley, *Modern Australian Documents 1939–1970*, vol. 2, Wren Publishing, Melbourne, 1973.

'U.S. alarm at leakage of Defence secrets', *Herald* (Melbourne), 8 March 1948, clippings package, NAA: A5954, 1956/6.

'Atomic bomb exploded in Monte Bello Islands', *West Australian*, 4 October 1952, reproduced in FK Crowley, *Modern Australian Documents 1939–1970*, vol. 2, Wren Publishing, Melbourne, 1973.

'His responsibility!', *Daily Mirror* (Sydney), 2 October 1953, p. 2.

'Atom bombs in our arid lands', *Sunday Herald*, 4 October 1953, p. 2.

'Atomic bombs, uranium and Australia's future', *SMH*, 8 October 1953, p. 2.

Staff Correspondent, 'Public safety and the Woomera tests', *SMH*, 9 October 1953, p. 2.

Staff Correspondent, 'Long-range risks in atomic tests?', *SMH*, 10 October 1953, p. 2.

'The atom's challenge to humanity', *SMH*, 16 October 1953, p. 2.

A 'Herald' Special Reporter who watched the atomic explosion on the Woomera Range from 15 miles away, 'Atom explosion success: dawn blast at Woomera', *SMH*, 16 October 1953, p. 1.

'Months of hard, lonely work paved way for atomic explosion', *SMH*, 16 October 1953, p. 4.

Special Correspondent in London, 'How an atomic bomb explodes', *SMH*, 16 October 1953, p. 2.

'Other "minor" tests', *SMH*, 28 October 1953, p. 1.

'Second A-blast successful: trial series ends', *SMH*, 28 October 1953, p. 1.

Special Correspondent, 'Australian atomic test site', *SMH*, 17 May 1955, p. 2.

'Labor opposes atomic tests', *Advertiser*, 21 June 1956, clippings package, NAA: A6456, R058/006.

'Threat to 3 towns', *Sun* (Sydney), 21 June 1956, p. 1.

'Misgivings over atomic tests', *The Age*, 22 June 1956, clippings package, NAA: A5954, 2167/6.

'Put method in atomic check', *Herald* (Melbourne), 22 June 1956, clippings package, NAA: A5954, 2167/6.

'Labor will stop weapons test', *Canberra Times*, 25 June 1956, clippings package, NAA: A6456, R058/006.

Staff Correspondent, 'Safety of A-weapon tests at Maralinga', *SMH*, 28 June 1956, p. 2.

'Latest on the bomb!', *Sun* (Sydney), 25 September 1956, p. 1.

'Freedom safeguard: more atomic tests', *Sun* (Sydney), 11 October 1956, p. 40.

'Reward for a good boy', *Sun-Herald*, 19 January 1958, clipping in biographical file for Howard Beale, National Library of Australia.

'Nuclear waste dump in SA: ex-RAAF man', *Advertiser*, 3 December 1976, p. 1.

'Nuclear dump "does exist" at Maralinga', *Advertiser*, 4 December 1976, p. 3.

Defence Reporter, 'Doubts cast on SA nuclear dump', *SMH*, 11 February 1977, p. 3.

'Nuclear waste could be stolen, govt told', *SMH*, 5 October 1978, p. 1.

'The usual four guard the plutonium', *SMH*, 7 October 1978, p. 1.

'Plutonium "unlikely"', sidebar to Andrew Kruger, 'Britain asked to take atomic waste', *SMH*, 10 October 1978, p. 1.

'Killen attacks Review report', *SMH*, 12 October 1978, p. 9.

'After Maralinga' (editorial), *Advertiser*, 17 April 1980, p. 5.

Obituary for Lord Penney, *The Times*, 6 March 1991.

ABC News South Australia, 12 June 1993, various bulletins.

New Scientist editorial comment, 12 June 1993, p. 3.

Obituary for Maurice Timbs, *The Australian*, 22 December 1994.

'Alan Nunn May, 91, pioneer in atomic spying for Soviets', *New York Times*, 25 January 2003.

'Spy's deathbed confession: atom physicist tells how secrets given to Soviet Union', *Guardian*, 27 January 2003.

NATIONAL ARCHIVES OF AUSTRALIA

NAA, series no. A816, Correspondence Files, Multiple Number Series [Classified 301], 1928–1962, particularly the following items:

3/301/539A, 'Operation Hurricane' Background Articles for Press etc, 1952.

10/301/128, Security of Defence Information – Confidential 'D' Notices to the Press – Main Policy File, 1947–1963.

10/301/129, Security of Defence Information – Confidential 'D' Notices to the Press – Australian 'D' Notices – File Number 1 (Covers D Notices 1 to 8), 1952–1955.

10/301/130, Security of Defence Information. Confidential 'D' Notices to the Press, 1947–1954.

10/301/131 & 10/301/132, Distribution of 'D' Notices by Secretary, Defence – Press and Broadcasting Committee, 1952.

NAA, series no. A1209, Correspondence Files, Annual Single Number Series [Classified] with Occasional C [Classified] Suffix, 1957–.

1957/5486, Security of Defence Information. 'D' Notices to the Press, 1950–1956.

NAA, series no. A5915, Whitlam Ministries – Cabinet Submissions, 1972–1975.

258, Proposed Australian Ionising Radiation Advisory Council – Decision 522, 1973.

NAA, series no. A5954, 'The Shedden Collection' [Records Collected by Sir Frederick Shedden during His Career with the Department of Defence and in Researching the History of Australian Defence Policy], Two Number Series, 1901–1971, particularly the following items:

1594/2, Defence, Press and Broadcasting Committee, 1937–1971.

1956/6, Security of Defence Information – Confidential 'D' Notices to the Press, 1947–1952.

2167/6, Atomic Tests – United Kingdom Tests at Monte Bello Island – May 1956. Maralinga – September 1956 [Newspaper Clippings], 1956.

NAA, series no. A6448, Royal Commission into British Nuclear Tests in Australia during the 1950s and 1960s – Transcripts of Proceedings, 1984–1985.

NAA, series no. A6449, Statements Received from United Kingdom Witnesses by the Royal Commission into British Nuclear Tests in Australia during the 1950s and 1960s, 1985.

NAA, series no. A6450, Statements Received from Australian Witnesses by the Royal Commission into British Nuclear Tests in Australia during the 1950s and 1960s, 1984–1985.

NAA, series no. A6454, 'Z Series' – [British] Atomic Weapons Research Establishment Reports, Single Number Series with Z or ZB Prefix, 1949–1985, particularly the following items:

Z5A, Atomic Weapons Research Establishment Report No T1/54 – Operation Hurricane Director's Report – Scientific Data Obtained at Operation Hurricane – November 1954, Copy 36, 1954.

ZB29, Atomic Weapons Research Establishment AWRE Report No T4/61 – Operation Vixen B1 [Sanitized Photocopy – Extracts of Introduction by DJ Collyer, NS (Nuclear Safety) Group Report by Major J McLean, DC (?Decontamination) Group Report by AE Oldbury], 1961.

NAA, series no. A6455, Exhibits Tendered before the Royal Commission into British Nuclear Tests in Australia during the 1950s and 1960, 1954–1985, particularly the following items:

RC371, Maralinga Experimental Programme [MEP] – Safety Statements for the Years 1959–62 and 1963 (Document 30–34 of Statement) – Presented 4/3/85 at London by Schofield, A, 1985.

RC386, Four Documents re Minor Trials – Maralinga Minor Trials in Relation to a Ban on Nuclear Testing – Letter Brooking, P to Sir W Cook re Minor Trials – Letter Brooking to AWRE re Minor Trials – Letter ADD/AWRE to Director re Minor Trials – Presented 6/3/85 at London by Pearce, N, 1985.

RC596, Royal Commission into British Nuclear Tests in Australia – File of Press Releases Presented to the Royal Commission in Sydney 1985, 1956–1957.

NAA, series no. A6456, Original Agency Records Transferred to the Royal Commission into British Nuclear Tests in Australia during the 1950s and 1960s, 'R' Series, 1952–1985, particularly the following items:

R021/001 PART 36, Department of Air – Statement by Mr Churchill in House of Commons on Monte Bello Atomic Test – [Letter Concerning Security of Information Regarding Atomic Test] – Substance of Communication Dated 18 October 1952 from High Commissioner for the United Kingdom, Canberra, 1952.

R023/003, Attorney-General's Department – [Draft of Publication] British Atomic Tests in Australia – Chronology of Events – 1950–1968 – Dr JL Symonds, 1984–1985.

R029/249, [Department of Supply] – Press Presentations at First Buffalo Test, Maralinga – Antler, 1957, 1956–1986.

R030/074 & R030/075, [Department of Supply] Maralinga Press and Publicity, 1956–1984.

R047/011, [Department of Defence] Press Statements – Atomic Tests in Australia, 1956–1985.

R047/022, [Department of Supply] Report on Maralinga Range – 1959 by GD Solomon (Range Commander), 1959–1985.

R055/011, [Department of Supply] Maralinga Atomic Proving Ground – Long Range Radiation Fall Out, 1957–1984.

R058/006 [Oliver Howard Beale, Minister for Department of Supply] Maralinga – TAG Hungerford Articles, 1955–1985.

R065/080, [Department of Prime Minister and Cabinet] Maralinga – Future Level of Activity, 1956–1985.

R069/032, [Department of Supply] AWTSC – Radiological Hazard of Plutonium, 1963–1985.

R075/002 [Department of Defence] Future Atomic Tests in Australia – Letter Beale to Menzies, 1953–1988.

R087/090, [Department of Supply] Operation Antler – 1957 Atomic Tests at Maralinga – Press publicity, 1957–1985.

R087/135, [Department of Supply] Maralinga – Press Visit – June 1956, 1955–1956.

R087/205, Series of Unrelated Files – VIP Trip – Visit Salisbury/Maralinga, 1950–1985.

R088/026, Loss of Captive Balloons from Maralinga 23–24 Sept 1960 – Report, 1960.

R091/003, Attendance of Australian Defence Scientific Adviser at Monte Bello, 1952.

R094/009, Maralinga – Safety Committee, 1950–1985.

R096/006, Professor Martin – Visit to Monte Bello, 1952–1953.

R105/001, Maralinga – SA, 1950–1985.

R107/005, Vixen B Trials – Correspondence, 1950–1985.

R120/111, Atomic Weapons Tests Safety Committee – Miscellaneous Papers, 1962–1966.

R120/137 PART 1, Atomic Weapons Tests Safety Committee, 1955-1962.

R124/025, Maralinga Atomic Tests – Security and Safety Precautions, 1955–1964.

R129/002, UKDRSS Maralinga Vixen 'B' (Taranaki), 1950–1985.

R145/011, Maralinga Project, 1950–1985.

R150/001, Maralinga Atomic Proving Ground – Minor Trials, 1958–1960.

R188/014, UK/Australia Use of Maralinga, 1950–1985.

NAA, series no. A12909, Second, Third, Fourth and Fifth Fraser Ministries – Cabinet Submissions (with Decisions), 1975–1983.

2605, Plutonium Buried near Maralinga Airfield – Decision 6812 (FAD), 1978.

OTHER PRIMARY SOURCES

Adams, CA, 'Co 60 Pellets Found at Maralinga', top-secret memorandum to Admiral PW Brooking, 12 August 1958, National Archives of the UK, DEFE 16/144.

Anderson, Ian, interview with unknown interviewee (ministerial staffer), 1993 (exact date unknown), Anderson files.

Anderson, Ian, interview on *Daybreak Show*, June 1993, Anderson files.

Anderson, Ian, interview on Tony Delroy's late night show, ABC Radio National, 10 June 1993, Anderson files.

Anderson, Ian, statement in support of application for a Michael Daley award, 29 September 1993, Anderson files.

Assistant Secretary, Research and Development, Department of Supply, internal memorandum to Secretary, 2 May 1963, Malone files.

Australia, House of Representatives, *Debates*, vol. 102, 1976, p. 3574.

Australia, House of Representatives, *Debates*, vol. 104, 1977, p. 880.

Australia, House of Representatives, *Debates*, 'Radioactive Material at Maralinga', ministerial statement by Mr Killen, 11 October 1978.

Australian Nuclear Veterans Association, South Australia & Maralinga and Monte Bello Islands Ex-Servicemen's Association, submissions, quoted in 'Protection was "inadequate"', *SMH*, 19 September 1995.

Australian Participants in British Nuclear Tests (Treatment) Bill 2006, bills digest no. 31, 2006–07, Parliament of Australia Parliamentary Library.

Australia, Senate, *Debates*, questions without notice, plutonium residues, Maralinga, 31 May 1984, p. 2227.

Beadell, Len, transcript of 'Too long in the bush', talk to Rotary Convention, Shepparton, Victoria, 1991, private recording.

Beale, Howard, Minister for Supply, 'Atomic Tests in Australia', top-secret Cabinet briefing document, submission no. 73, Canberra, 11 August 1954, Malone files.

Beale, Howard, Minister for Supply, memorandum, Canberra, 4 May 1955, National Archives of the UK, FCO 1 358/2099.

Beale, Howard, letter to Paul Hasluck, Minister for Territories, 1 May 1957, Malone files.

Brown, HJ, Woomera, confidential cable re press reports about Aboriginal children being removed from their parents, to O'Connor, Department of Supply, 21 November 1956, Malone files.

Brown, HJ, Controller, Weapons Research Establishment, Salisbury, cable to EL Cook, Department of Supply, 2 December 1957, Malone files.

Burns, Peter & Williams, Geoff, tape recording of interview with Ian Anderson, 1993 (exact date unknown), Anderson files.

Burns, Peter & Williams, Geoff, interview with author, ARPANSA, Melbourne, 15 April 2004.

Butement, Alan, Chief Scientist, memorandum to Controller, Weapons Research Establishment, 8 June 1956, Malone files.

Clark, Emeritus Professor AM, Chair, AIRAC, letter to Barry Cohen, Minister for Home Affairs and Environment, 13 August 1984, Malone files.

Clarke QC, John, 'British atomic tests', in Department of Veteran Affairs, *Review of Veterans' Entitlements*, Canberra, 6 January 2003, Chapter 16.

Commission of Enquiry into the Australian Secret Intelligence Service, 'The D-notice system', in *Report on the Secret Intelligence Service*, public edn, AGPS, Canberra, March 1995, Chapter 11.

Crouch, PC, Robotham, FJP & Williams, GA, 'Submission to the Senate Foreign Affairs, Defence and Trade Committee on the Australian Participants in British Nuclear Tests (Treatment) Bill 2006; and the Australian Participants in British Nuclear Tests (Treatment) (Consequently Amendments and Transitional Provisions) Bill 2006', 25 October 2006.

Davoren, Patrick, personal correspondence with author, 13 May 2004.

Department of Education, Science and Training, 'Rehabilitation of Former Nuclear Tests Sites at Emu and Maralinga (Australia) 2003' (MARTAC Report), Commonwealth of Australia, Canberra, 2002.

Department of Supply, notes of interdepartmental committee meeting to consider decision no. 69 (PM) Cabinet minute dated 26 August 1954, 6 September 1954, Malone files.

Dudgeon, Henry, confidential cable to Foreign and Commonwealth Office, 3 November 1978, National Archives of the UK, FCO 1/20.

Durie, R, Acting Secretary to Cabinet, Cabinet minute, decision no. 69, Prime Minister's Committee, Canberra, 26 August 1954, Malone files.

Expert Committee on the Review of Data on Atmospheric Fallout Arising from British Nuclear Tests in Australia, *Report* (CB Kerr, Chair) (Kerr Report), 31 May 1984, Malone files.

Fadden, Arthur, 'UK Atomic Test Woomera – D-notice to the Press', letter to Press Chiefs, 26 June 1953, in JL Symonds, *A History of British Atomic Tests in Australia*, AGPS, Canberra, 1985, p. 151.

Bibliography

Foulkes, JN & Thompson, DS, 'A Biological Survey of the Maralinga Tjuratja Lands, South Australia, 2001–2007', Science Resource Centre, Information, Science and Technology Directorate, Department for Environment and Heritage, South Australia, December 2008.

Fraser, Malcolm, Prime Minister of Australia, letter to South Australian Premier Don Dunstan, 12 February 1979, Malone files.

Garland, RV, Minister for Supply, letter to P Howson, Minister for the Environment, Aborigines and the Arts, Canberra, 2 February 1972, Malone files.

Hamilton, Archie, House of Commons, *Debates*, vol. 222, 1 April 1993, cc. 650–651, <hansard.millbanksystems.com/commons/1993/apr/01/atomic-test-site-south -australia#column_650>. Accessed 23 March 2016.

Harman, JJ, Divisional Superintendent, Eastern Division, Western Australian Department of Native Welfare, cable to Secretary, Department of Supply, Melbourne, 1 May 1963, Malone files.

Kerr Report, see Expert Committee on the Review of Data on Atmospheric Fallout Arising from British Nuclear Tests in Australia.

Lokan, KH, Head of ARL, evidence to Joint Committee on Public Works, 23 February 1995, in minutes of evidence relating to Maralinga Rehabilitation Project, Parliament of the Commonwealth of Australia, 1995.

Lokan, KH & Williams, GA, 'Maralinga – Radiological Safety for APS Personnel', first prepared June 1988, revised January 1995, attachment to PC Crouch, FJP Robotham & GA Williams, 'Submission to the Senate Foreign Affairs, Defence and Trade Committee on the Australian Participants in British Nuclear Tests (Treatment) Bill 2006; and the Australian Participants in British Nuclear Tests (Treatment) (Consequently Amendments and Transitional Provisions) Bill 2006', 25 October 2006.

Macaulay, R, 'Report on Mr. L. Beadell', letter to Superintendent, Woomera, 17 April 1957, Malone files.

Macaulay, R, 'Shell Lakes–Boundary Dam Patrol May–June 1959', letter to Commissioner, Native Welfare, Perth, Western Australia, 12 January 1960, Malone files.

Macaulay, R, 'Native Reserves of South-Western Central Australia – Their Future and Department of Supply Interest', report for Weapons Research Establishment, Woomera, 28 March 1960, Malone files.

Macaulay, R, report to Superintendent, Woomera, 16 October 1963, Malone files.

MacDougall, WB, reports to Superintendent, Woomera, 31 January 1952, 7 October 1953, 10 December 1953, July 1955, 11 October 1955, 16 January 1956, Malone files.

MacDougall, WB, 'Proposed Survey of Tribal Aborigines Living near the South Eastern Portion of the Central Reserve', sent to Superintendent, Woomera, 3 March 1952, Malone files.

MacDougall, WB, 'Detailed Survey of the Jangkuntjara Tribe – Their Traditional Tribal Country and Ceremonial Grounds', sent to Superintendent, Woomera, 2 March 1953, Malone files.

MacDougall, WB, 'Patrol at Ooldea', minute paper for Department of Supply, 5 February 1954, Malone files.

MacDougall, WB, letter to Secretary, Aborigines Protection Board, Adelaide, November 1954, Malone files.

MacDougall, WB, Native Patrol Officer, letter to Superintendent, Woomera,
4 November 1955, Malone files.

MacDougall, WB, 'Report on Patrol to Central Reserve Including That Area
Recently Excised for the Use of the South West Mining Company', letter to
Commissioner, Native Welfare, Western Australia, 3 August 1956, Malone files.

MacDougall, WB, 'Report on Patrol in Connection with Maralinga Tests 3rd
September, to 13th October, 1957', report to Superintendent, Woomera,
18 October 1957, Malone files.

MacDougall, WB, 'Report on Patrol to Rawlinson Ranges 6th to 30th January, 1958',
report to Superintendent, Woomera, 25 February 1958, Malone files.

Maralinga Peace Officer Guard, 'Native Movement – Maralinga', letter to Deputy
Superintending Peace Officer, Adelaide, 16 May 1957, Malone files.

Maralinga Rehabilitation Project, Parliament of the Commonwealth of Australia,
minutes of evidence, 1995.

MARTAC Report, see Department of Education, Science and Training.

Ministry of Defence (UK), draft telegram prepared 29 April 1955 and received by the
Foreign and Commonwealth Office 4 May 1955, National Archives of the UK,
FCO 1/8.

Moroney, John, cable to HJ Brown, Controller, Weapons Research Establishment,
Salisbury, 2 December 1957, Malone files.

Moroney, John, letter to Patrick Davoren, 28 November 1991.

Moroney, John, 'Aide Memoire on Health Control and Surveillance in the Nuclear
Tests in Australia', unpublished briefing document, ARL, 24 December 1992.

Moroney, John, 'Aide Memoire on the Nuclear Tests in Australia in the Context of the
British Weapons Development Program', unpublished briefing document, ARL,
24 December 1992.

Moroney, John, annotated draft of Ian Anderson's *New Scientist* story, unpublished,
2 June 1993, Anderson files.

Moroney, John, 'Statement on the Disagreement in the Plutonium Data at Maralinga
in the Discussion with the Australian Editor of *New Scientist*', briefing for Geoff
Williams & Pat Davoren, 3 June 1993.

O'Connor, FA, 'Atomic Tests in Australia – Report of Inter-Departmental Committee',
briefing note for Minister for Supply, 29 September 1954, Malone files.

Pearce, Noah, 'Final Report on Residual Radioactive Contamination of the Maralinga
Range and the Emu Site', AWRE report no. 01–16/68 (Pearce Report), UK
Atomic Energy Authority, January 1968. Edited version released by Australian
Department of National Development, May 1979. Full version tabled by
Australian Senate, May 1984.

Pearce Report, see Pearce, Noah.

Pilgrim, Roy, 'Vixen B – Firing Safety Circuit', memorandum to AWRE Deputy
Director EF Newley, 23 August 1960, National Archives of the UK, ES 1/962.

Pilgrim, Roy, letter to Ernest Titterton, 11 October 1962, National Archives of the UK,
DEFE 16/854.

Powell, RR, UK Ministry of Defence, letter to JM Wilson, UK Ministry of Supply,
9 February 1955, National Archives of the UK, FCO 1/8.

'Rehabilitation of the former nuclear tests sites at Maralinga and Emu', in *Annual
Report of the Chief Executive Officer of ARPANSA, 1998–1999*, ARPANSA,
Melbourne, 1999, pp. 35–38.

Bibliography

Richardson, JF, Australian Health Physics Representative, 'Report to the Australian Atomic Weapons Tests Safety Committee on Visits to Maralinga during Operation Brumbie [sic]', 19 July 1967, Malone files.

Royal Commission into British Nuclear Tests in Australia, *Conclusions and Recommendations* (JR McClelland, J Fitch, WJA Jones), AGPS, Canberra, 1985.

Royal Commission into British Nuclear Tests in Australia, *Report* (JR McClelland, President), 2 vols, AGPS, Canberra, 1985.

Secretary, Aborigines Protection Board, letters to Mr WB MacDougall, LRWE Range, Woomera, 15 December 1953 & 14 January 1955, Malone files.

Secretary of State for Commonwealth Relations, top-secret telegram to High Commissioner for the UK, Australia, 2 April 1955, National Archives of the UK, FCO/1/8.

Security Officer, Maralinga (name redacted), 'Control of Aborigines – Maralinga', letter to Range Commander, Maralinga, 25 July 1957, Malone files.

Smith, F, 'Report on Natives at PomPom – 14 May 57', letter to Harry Turner, Health Physics Adviser, 15 May 1957, Malone files.

Stewart, K, 'Vixen "B" 1960: Nuclear Safety Group Experimental Plan', 8 July 1960, National Archives of the UK, ES 1/962.

Thorn, Robert N & Westervelt, Donald R, 'Hydronuclear Experiments', report for Los Alamos National Laboratories, LA-10902-MS, February 1987.

Toohey, Brian, personal correspondence with author, 31 December 2009.

Turner, OH, 'Health Physics Report on Natives at Pom Pom – 14 May 57', letter to Range Commander, Maralinga, 18 May 1957, Malone files.

United Nations Office for Disarmament Affairs, 'Treaty on the Non-Proliferation of Nuclear Weapons: text of the treaty', <disarmament.un.org/treaties/t/npt/text>. Accessed 28 February 2016.

Walsh, Peter, statement, Senate, 4 May 1984, reproduced in *Australian Foreign Affairs Record*, vol. 55, no. 5, May 1984, p. 486.

Walsh, Peter, media release, 15 May 1984, reproduced in *Australian Foreign Affairs Record*, vol. 55, no. 5, May 1984, p. 547.

Walsh, Peter, statement, Senate, 7 June 1984, reproduced in *Australian Foreign Affairs Record*, vol. 55, no. 6, June 1984, p. 636.

Walsh, Peter, letter to Howard Conkey, *Canberra Times*, 28 June 1984.

Walsh, Peter, media release, 5 July 1984, reproduced in *Australian Foreign Affairs Record*, vol. 55, no. 7, July 1984, p. 732.

Webb, Jeremy, *New Scientist*, personal correspondence with author, 25 May 2004.

Williams, Geoff, ARPANSA, personal correspondence with author, 27 July 2010 & 23 February 2016.

Wills, HA, Chief Executive Officer, Maralinga Committee, 'Aboriginal Population of South Australia', letter to HJ Brown, Controller, Weapons Research Establishment, Salisbury, 6 June 1956, Malone files.

SECONDARY SOURCES

Acaster, Ray, 'Worlds apart: atom bombs and traditional land use in South Australia', *Limina*, vol. 1, 1995.

Arnold, Lorna, *A Very Special Relationship: British Atomic Weapon Trials in Australia*, Her Majesty's Stationery Office, London, 1987.

Arnold, Lorna & Smith, Mark, *Britain, Australia and the Bomb: The Nuclear Tests and Their Aftermath*, 2nd edn, Palgrave Macmillan, New York, 2006.

Arvanitakis, James, 'Staging Maralinga and looking for community (or why we must desire community before we can find it)', *Research in Drama Education*, vol. 13, no. 3, pp. 295–306.

Aylen, Jonathan, 'First waltz: development and deployment of Blue Danube, Britain's post-war atomic bomb', *International Journal for the History of Engineering and Technology*, vol. 85, no. 1, January 2015, pp. 31–59.

Barnaby, Frank, 'The management of radioactive wastes and the disposal of plutonium', paper at Medical Association for Prevention of War Conference, 2000.

Beadell, Len, *Blast the Bush*, Rigby Limited, Adelaide, 1967.

Beale, Howard, *This Inch of Time: Memoirs of Politics and Diplomacy*, Melbourne University Press, Melbourne, 1977.

Bernstein, Jeremy, *Plutonium: A History of the World's Most Dangerous Element*, UNSW Press, Sydney, 2001.

Blakeway, Denys & Lloyd-Roberts, Sue, *Fields of Thunder: Testing Britain's Bomb*, George Allen & Unwin, London, 1985.

Brown, Paul, 'British nuclear testing in Australia: performing the Maralinga experiment through verbatim theatre', *Journal and Proceedings of the Royal Society of New South Wales*, vol. 139, 2006.

Cawte, Alice, *Atomic Australia 1944–1990*, UNSW Press, Kensington, 1992.

Clearwater, John & O'Brien, David, '"Oh lucky Canada": radioactive polar bears; the proposed testing of British nuclear weapons in Canada', *Bulletin of the Atomic Scientists*, July–August 2003.

Cross, Roger, *Fallout: Hedley Marston and the British Bomb Tests in Australia*, Wakefield Press, Kent Town, 2001.

Cross, Roger & Hudson, Avon, *Beyond Belief: The British Bomb Tests; Australia's Veterans Speak Out*, Wakefield Press, Kent Town, 2005.

Crowley, FK, *Modern Australian Documents 1939–1970*, vol. 2, Wren Publishing, Melbourne, 1973.

Cryle, Denis, 'Rousing the British-speaking world: Australian newspaper proprietors and freedom of the press, 1940–1950', ARC-funded project, Central Queensland University, paper at Australian Media Traditions Conference, 2007.

Curr, EM, 'Port Essington', in *Australian Race: Its Origin, Languages, Customs, Place of Landing in Australia, and the Routes by Which It Spread Itself over That Continent*, John Ferres, Government Printer, 1887.

Curry Jansen, Sue & Martin, Brian, 'Exposing and opposing censorship: backfire dynamics in freedom-of-speech struggles', *Pacific Journalism Review*, vol. 10, no. 1, April 2004, pp. 29–45.

Dousset, Laurent, 'Politics and demography in a contact situation: the establishment of the Giles Meteorological Station in the Rawlinson Ranges, West Australia', *Aboriginal History*, vol. 26, 2002.

Duguid, Charles, 'The Central Aborigines Reserve', presidential address to Aborigines Advancement League of South Australia, 21 October 1957.

Fairley, Douglas, 'D notices, official secrets and the law', *Oxford Journal of Legal Studies*, vol. 10, no. 3, Autumn 1990, p. 432.

Fitzgerald, EM, 'Allison, Attlee and the bomb: views on the 1947 British decision to build an atom bomb', *RUSI Journal*, vol. 122, issue 1, 1977.

Bibliography

Goodman, Michael S, 'The grandfather of the hydrogen bomb? Anglo-American intelligence and Klaus Fuchs', *Historical Studies in the Physical and Biological Sciences*, vol. 34, no. 1, 2003, pp. 1–22.

Gowing, Margaret, *Britain and Atomic Energy 1939–1945*, Macmillan, London, 1964.

Gowing, Margaret, 'James Chadwick and the atomic bomb', *Notes and Records of the Royal Society of London*, vol. 47, no. 1, January 1993.

Gowing, Margaret with Arnold, Lorna, *Independence and Deterrence: Britain and Atomic Energy, 1945–1952*, Palgrave Macmillan, 1974.

Grabosky, PN, 'A toxic legacy: British nuclear weapons tests', in *Wayward Governance: Illegality and Its Control in the Public Sector*, Australian Institute of Criminology, Canberra, 1989, Chapter 16.

Gregory, Shaun, *The Hidden Cost of Deterrence: Nuclear Weapons Accidents*, Brassey's (UK), London, 1990.

Hennessy, Peter, 'Cabinets and the bomb', Inaugural Michael Quinlan Lecture, UK House of Lords, 2 February 2011.

Ison, Erin, 'Bombs, "reds under the bed" and the media: the Menzies government's manipulation of public opinion, 1949–1957', *Humanity*, University of Newcastle, 2008.

Kendall, GM, Muirhead, CR, Darby, SC, Doll, R, Arnold, L & O'Hagan, JA, 'Epidemiological studies of UK test veterans: I. General description', *Journal of Radiological Protection*, vol. 24, 27 August 2004, p. 203.

Leonard, Zeb, 'Tampering with history: varied understandings of Operation Mosaic', *Journal of Australian Studies*, vol. 38, no. 2, May 2014, p. 219.

Macintyre, Stuart, *A Concise History of Australia*, 2nd edn, Cambridge University Press, Port Melbourne, 2004.

Maher, Laurence W, 'National security and mass media self-censorship: the origins, disclosure, decline and revival of the Australian D-notice system', *Australian Journal of Legal History*, vol. 3, 1997.

Maloney, Sean M, *Learning to Love the Bomb: Canada's Nuclear Weapons during the Cold War*, Potomac Books, 2007.

Martin, AW, *Robert Menzies: A Life*, vol. 2, *1944–1978*, Melbourne University Press, Melbourne, 1999.

Mattingley, Christobel, 'Atom bombs before Aborigines: Maralinga', in Christobel Mattingley & Ken Hampton, *Survival in Our Own Land: 'Aboriginal' Experiences in 'South Australia' since 1836*, ALDAA in association with Hodder & Stoughton, Adelaide, 1988.

Mazel, Odette, 'Returning Parna Wiru: restitution of the Maralinga lands to traditional owners in South Australia', in Marcia Langton (ed.), *Settling with Indigenous People: Modern Treaty and Agreement Making*, Federation Press, 2006.

McClelland, James, *Stirring the Possum: A Political Autobiography*, Penguin, Ringwood, 1988.

McClelland, James, transcript of interview with Robin Hughes, Australian Biography Project, 25 January 1995, <www.australianbiography.gov.au/subjects/mcclelland/interview6.html>. Accessed 20 November 2015.

McClellan, Peter, 'Who is telling the truth? Psychology, common sense and the law', Local Courts of New South Wales Annual Conference, August 2006.

McKnight, David, *Australian Spies and Their Secrets*, Allen & Unwin, Sydney, 1994.

Menzies, Robert, *The Measure of the Years*, Cassell Australia, Melbourne, 1970.

Michel, Dieter, 'Villains, victims and heroes: contested memory and the British nuclear tests in Australia', *Journal of Australian Studies*, no. 80, 2004.

Milliken, Robert, *No Conceivable Injury*, Penguin, Ringwood, 1986.

Newton, JO, 'Ernest William Titterton 1916–1990', *Historical Records of Australian Science*, vol. 9, no. 2, 1992.

Norris, Robert Standish, *Questions on the British H-Bomb*, Natural Resources Defense Council, 22 June 1992.

Ophel, Trevor & Jenkin, John, *Fire in the Belly: The First 50 Years of the Pioneer School at the ANU*, ANU, Canberra, 1996.

Parkinson, Alan, 'The Maralinga rehabilitation project: final report', *Journal of Medicine, Conflict and Survival*, vol. 20, 2004.

Parkinson, Alan, *Maralinga: Australia's Nuclear Waste Cover-Up*, ABC Books, Sydney, 2007.

Paterson, Robert H, *Britain's Strategic Nuclear Deterrent: From before the V-Bomber to beyond Trident*, Routledge, 2012.

Rankin, Scott, 'Ngapartji Ngapartji', in *Namatjira / Ngapartji Ngapartji*, Currency Press, Sydney, 2012.

Rankin, Scott, 'Ngapartji Ngapartji', Wikipedia, <en.wikipedia.org/wiki/Ngapartji_Ngapartji>. Accessed 26 March 2016.

Sadler, Pauline, 'The D-notice system', *Press Council News*, vol. 12, no. 2, May 2000, <www.presscouncil.org.au/pcsite/apcnews/may00/dnote.html>.

Sadler, Pauline, *National Security and the D-notice System*, Dartmouth Publishing Company, Aldershot, 2001.

Schedvin, CB, *Shaping Science and Industry: A History of Australia's Council for Scientific and Industrial Research, 1926–49*, Allen & Unwin, Sydney, 1987.

Schrafstetter, Susanna, '"Loquacious … and pointless as ever"? Britain, the United States and the United Nations negotiations on international control of nuclear energy, 1945–48', *Contemporary British History*, vol. 16, no. 4, 2002, pp. 87–108.

Seldon, Robert, Lawrence Livermore Laboratory, quoted in Frank Barnaby, Plutonium and Radioactive Waste Management Policy Consultant, 'The management of radioactive wastes and the disposal of plutonium', paper at Medical Association for Prevention of War Conference, 2000.

Sherratt, Tim, 'A political inconvenience: Australian scientists at the British atomic weapons tests, 1952–53', *Historical Records of Australian Science*, vol. 6, no. 2, December 1985.

Smith, Joan, *Clouds of Deceit: The Deadly Legacy of Britain's Bomb Tests*, Faber & Faber, London, 1985.

Spinardi, Graham, 'Aldermaston and British nuclear weapons development: testing the "Zuckerman thesis"', *Social Studies of Science*, vol. 27, no. 4, 1997, pp. 547–582.

Summers, Anne, *Gamble for Power: How Bob Hawke Beat Malcolm Fraser; The 1983 Federal Election*, Thomas Nelson Australia, Melbourne, 1983.

Symonds, JL, *A History of British Atomic Tests in Australia*, AGPS, Canberra, 1985.

Tame, Adrian & Robotham, FP, *Maralinga: British A-Bomb Australian Legacy*, Dominion Press, Melbourne, 1982.

Titterton, EW, *Facing the Atomic Future*, Macmillan, London, 1956.

Toohey, Brian & Wilkinson, Marian, *The Book of Leaks: Exposés in Defence of the Public's Right to Know*, Angus and Robertson, North Ryde, 1987.

Turner, Graeme, 'Of rocks and hard places: the colonized, the national and Australian cultural studies', *Cultural Studies*, vol. 6, no. 3, 1992.

Bibliography

Uren, Tom, *Straight Left*, Vintage, Milsons Point, 1995.

Walker, Frank, *Maralinga: The Chilling Exposé of Our Secret Nuclear Shame and Betrayal of Our Troops and Country*, Hachette Australia, Sydney, 2014.

Walker, John R, *British Nuclear Weapons and the Test Ban 1954–1973*, Ashgate, Surrey, 2010.

Walsh, Peter, *Confessions of a Failed Finance Minister*, Random House Australia, Milsons Point, 1995.

Wick, OJ, *Plutonium Handbook: A Guide to the Technology*, American Nuclear Society, Grange Park, 1980.

Williams, Geoff & O'Brien, Richard, appraisal of *The Black Mist and Its Aftermath – Oral Histories by Lallie Lennon*, ARPANSA, Melbourne, 20 April 2010.

Woollett, SM, 'Rehabilitating Maralinga', *Clean Air and Environmental Quality*, vol. 37, issue 4, November 2003.

Wright, RVS, 'Bates, Daisy May (1863–1951)', *Australian Dictionary of Biography*, vol. 7, Melbourne University Press, Melbourne, 1979.

Yalata and Oak Valley Communities with Mattingley, Christobel, *Maralinga: The Anangu Story*, Allen & Unwin, Sydney, 2009.

WEBSITES

Allen, Christian, 'Atom spy Klaus Fuchs jailed', *History Today*, LookSmart, 2000, <www.historytoday.com/christian-allen/atom-spy-klaus-fuchs-jailed>. Accessed 19 November 2015.

'Broken arrows: nuclear weapons accidents', Atomic Archive, <www.atomicarchive. com/Almanac/Brokenarrows_static.shtml. Accessed 13 March 2016.

Cathcart, Brian, 'Obituary: Theodore Hall', *Independent*, 12 November 1999, <www. independent.co.uk/arts-entertainment/obituary-theodore-hall-1125267.html>. Accessed 9 December 2010.

Lewis, Richard, Morris, Annette & Oliphant, Ken, 'Tort personal injury claims statistics: is there a compensation culture in the United Kingdom?', Cardiff Law School, Cardiff University, <www.law.cardiff.ac.uk/research/pubs/repository/1445.pdf>. Accessed 30 November 2010.

Maher, Laurence W, 'H.V. Evatt and the Petrov defection: a lawyer's interpretation', *Australian Society of Labour History*, 2007, <www.historycooperative.org/ proceedings/asslh2/maher.html>. Accessed 31 December 2009.

'Maralinga lament', Australian Nexus, <www.ausnexus.co.uk/events/maralinga -lament/#.VlERyUYXgQ0>. Accessed 22 November 2015.

Maralinga Tours, <www.maralingatours.com.au>. Accessed 22 November 2015.

Meade, Roger & Meade, Linda (eds), 'Klaus Fuchs: the second confession', Los Alamos National Laboratory Historical Document, 2014, <permalink.lanl.gov/object/ tr?what=info:lanl-repo/lareport/LA-UR-14-27960>. Accessed 20 July 2015.

'The mushroom cloud', Atomic Archive, <www.atomicarchive.com/Effects/effects9. shtml>. Accessed 27 September 2015.

'Noah Pearce', Geni, <www.geni.com/people/Noah-Pearce/6000000000836708056>. Accessed 1 November 2015.

NTD Resource Center, 'Stochastic effects', <www.nde-ed.org/EducationResources/ CommunityCollege/RadiationSafety/biological/stochastic/stochastic.htm>. Accessed 27 March 2016.

'Nuclear colonialism', Healing Ourselves and Mother Earth, <www.h-o-m-e.org/nuclear-colonialism.html>. Accessed 23 August 2014.

Nuclear Futures, <nuclearfutures.org>. Accessed 22 November 2015.

Oppenheimer, J Robert, 'Now I am become death', video recording, Atomic Archive, <www.atomicarchive.com/Movies/Movie8.shtml>. Accessed June 2015.

'Plutonium and Aldermaston – an historical account', Federation of American Scientists, 2000, <fas.org/news/uk/000414-uk2.htm>. Accessed 19 November 2015.

Rees, Margaret, 'Documents confirm soldiers were exposed to nuclear tests in Australia', World Socialist, 9 July 2001, <www.wsws.org/articles/2001/jul2001/mara-j09.shtml>. Accessed 18 December 2009.

Resture, Jane, 'Return to Maralinga', Jane Resture's Oceania Page, 2012, <www.janesoceania.com/christmas_about1/index.htm>. Accessed 23 August 2014.

Sublette, Carey, 'Cobalt bombs and other salted bombs', Nuclear Weapon Archive, <nuclearweaponarchive.org/Nwfaq/Nfaq1.html#nfaq1.6>. Accessed 13 October 2015.

Truman, Harry S, 'Statement announcing the atomic bombing of Hiroshima, August 6, 1945', American Experience, <www.pbs.org/wgbh/americanexperience/features/primary-resources/truman-hiroshima>. Accessed 11 October 2015.

WikiLeaks, 'United Kingdom atomic weapons program: the full Penney report (1947)', <wikileaks.org/wiki/United_Kingdom_atomic_weapons_program:_The_full_Penney_Report_%281947%29>. Accessed 20 July 2015.

WikiLeaks, 'How Britain got the bomb: Britain's "Oppenheimer", William G. Penney and his secret report on what Britain needed to do to "get the bomb"', <wikileaks.org/wiki/How_Britain_got_the_bomb>. Accessed 19 November 2015.

Index

Page numbers followed by 'g' indicate glossary entries.

Index